Forests in Peril

Forests in Peril

Tracking Deciduous Trees
from Ice-Age Refuges into the Greenhouse World

Hazel R. Delcourt

Department of Ecology and Evolutionary Biology
University of Tennessee, Knoxville

The McDonald & Woodward Publishing Company
Blacksburg, Virginia

The McDonald & Woodward Publishing Company
Blacksburg, Virginia

Forests in Peril : Tracking Deciduous Trees from Ice-Age Refuges into the Greenhouse World

© 2002 by Hazel R. Delcourt

All rights reserved. First printing September 2002
Printed in the United States of America by
McNaughton & Gunn, Inc., Saline, MI

12 11 10 9 8 7 6 5 4 3 2   10 9 8 7 6 5 4 3 2 1

Library of Congress Cataloging-in-Publication Data
Delcourt, Hazel R.
    Forests in peril : tracking deciduous trees from ice-age refuges into the greenhouse world / Hazel R. Delcourt.
        p. cm.
Includes bibliographical references (p.   ).
    ISBN 0-939923-89-0 (alk. paper)
    1. Forests and forestry--East (U.S.) 2. Forests and forestry--Canada, Eastern. 3. Forest ecology--East (U.S.) 4. Forest ecology--Canada, Eastern. 5. Forests and forestry--East (U.S.). 6. Forests and forestry--Canada, Eastern--History. 7. Forest ecology--East (U.S.)--History. 8. Forest ecology--Canada, Eastern--History. I. Title.
    SD144.A112 D45 2002
    333.75'0974--dc21
                                                                    2002008793

Contents

To Paul,

my inspiration,
life companion,
and logistical support specialist

Acknowledgments

I owe a debt of appreciation to my academic mentors, who inspired me either directly through their tutoring or indirectly through their written word. In alphabetical order, I wish to thank the following for their guidance during formative stages of my academic career: Emma Lucy Braun, Clair Brown, Edward Cushing, Margaret Davis, Edward Deevey, Jim Kane, Ronald Kapp, Katherine Keever, Estella Leopold, Paul Martin, Milton Newton, Elsie Quarterman, William Watts, Donald Whitehead, and Herbert Wright, Jr.

I also thank the following colleagues who have collaborated in Quaternary field and lab research over the years: David Anderson, Jefferson Chapman, Kristen Gremillion, Cecil Ison, Gary Larsen, Steven Lund, Gerald Lang, Mark Lynott, Rex Mann, Jerry McDonald, Dan and Phyllis Morse, J. Dan Pittillo, Jim and Cynthia Price, Douglas Rossman, Bill Sharp, Roger Saucier, Allen Solomon, and Thompson Webb III.

I extend my thanks to each graduate student who has completed a thesis or dissertation with Paul and me based on study sites in the southeastern United States: Patricia Cridlebaugh, Jean Davidson, Peter Larabee, P. Daniel Royall, David Shafer, E. Newman Smith, Jr., Robert Tolliver, and Gary Wilkins.

Forests in Peril

Introduction:

Finding a Window into the Past

Forests of the Ozark highlands of southeastern Missouri can seem remote, located far from interstate highways and cities. Logging roads, many unpaved, connect small outposts of civilization such as Doniphan and Eminence. The town motto of Eminence, Missouri, is "Stay for a day or a lifetime." The summer we cored Cupola Pond, it seemed as if we had lived in Eminence for most of our lives.

The first day we saw Cupola Pond, I remember wiping the perspiration off my face with a red handkerchief that still smelled freshly laundered, and then tying it in back of my head, behind my long, single braid in "babushka" fashion. Even in the early morning, the June sun was fierce, as we rumbled down a back road in our non-air-conditioned station wagon. I picked up the topographic map, trying to judge our progress by matching map symbols for churches, orchards, and stream crossings with the frame buildings, forest patches, and wetlands that we were passing. As long as we kept a steady speed, I could usually decipher our location.

My husband and teammate, Paul, was driving. Despite the weather, he wore his usual heavy field garb: an olive-green long-sleeved shirt, pants, and high-topped waffle-lug-bottomed boots. His unkempt black beard and thick, horn-rimmed glasses framed a face topped by a dirty white terrycloth sweatband and sweat-stained green baseball hat. I told him he resembled a character straight from a grade B movie about third-world revolutionaries. He grimaced as he said, "Wearing full field gear hardens you to field work."

It made me feel uncomfortable just to look at Paul. I recall that on my first day in the field that season, I had been overdressed in heavy bib overalls and had nearly fainted in the process of climbing up a stream bank from an archaeological site where our friends Jim and Cynthia Price, of the southwestern Missouri archaeological field station in Naylor, Missouri, were finding relics of earlier life along the Current River. "Look, the lock from an old jailhouse door," Cynthia had exclaimed, holding up a rusted plate of metal in front of me while I held my dizzy head in my hands. After that day, I wore only a tee shirt and jeans into the field and I was able to adjust to the summer heat and humidity.

On the way to Cupola Pond, our graduate student, Newman Smith, who was usually cheerful and talkative, sat quietly in the back seat, staring out the window. Newman was a clean-cut, solid man, a little older than the average graduate student. He had been stationed in Germany when he was in the US Army, and for a time after that he had worked as a "repo man" specializing in reclaiming eighteen-wheel trucks in a rural part of Kentucky. Eventually he returned to college and became a geologist. Newman's good-natured attitude and ever-present sense of humor had sustained him in his previous endeavors. But the deeper into the Ozark highlands we trekked, the less certain he seemed to be about his chosen thesis topic. Would we find the right study site for him, one that would give us a window into the past?

As we drove the logging roads through Shannon, Carter, and Ripley counties, towing a rented trailer with our equipment, Paul kept us alert with stories about previous failed expeditions to remote Ozark highlands ponds. Although sinkholes and natural springs abound in southeastern Missouri — places with names like Round Spring and Alley Spring that attract tourists on summer vacations — we were no less than the fifth group of paleoecologists to go looking in the Ozark highlands for a deep sediment record. Our requirements were exacting. We couldn't use a site like Round Spring that was a conduit for fast outflow of copious amounts of spring water that would flush away any delicate plant remains that might have fallen into the sinkhole during its history. Instead, we needed to find a quiet pond that would contain old, deep, undisturbed clays and silts, sediments that would yield paleobotanical clues about what the environment of the Ozark highlands was like many thousands of years ago. Previous expeditions in search of the prehistory of the Ozark highlands landscape had failed, in part because sites that had looked promising on topographic maps were too remote and inaccessible. Many of the sinkholes were of very recent origin and did not offer a window into the remote past.

But none of the previous expeditions had journeyed to Cupola Pond. The Prices knew that its name dated back to the time when the area was settled by European American pioneers, some of whom were Jim's direct ancestors. No one we asked, however, knew why the pond was called "cupola." Other sinkholes in the region have names like Indian Pond and Tupelo Gum Pond. Tupelo gum (*Nyssa aquatica*) is a kind of tree that grows in permanently standing water around the margins of circular sinkholes. We speculated that cupola might be a corruption of the word tupelo. But it was also possible that the pioneers imagined that the trees towering above this small pond looked and functioned like the cupola atop a house — a natural ventilation system that would draw hot air off and let cool air sink down.

Wherever the name came from, to us Cupola Pond held the prospect of becoming a window into the past, a portal into a world undoubtedly very different from that of today, but one about which we could only speculate without fossil evidence. Jim and Cynthia had noticed that the pond held water year-around. This was critical to our work because ponds that dry out seasonally have their surface sediments exposed to the air, which causes the deterioration of their fossil contents over time. Forest Service biologists had identified rare plants growing near Cupola Pond, which told us that the pond probably had not been much disturbed by recent human activities. Because of its uniqueness and its naturalness, this pond and its surrounding forests had been designated a National Natural Landmark. This designation gave this site an official status that further protected it from timbering, draining, and other human activities that would disturb the pond, its nearby plant and animal populations, and even the mud beneath its surface. Such protection was vital, for if the sinkhole in which the pond formed had been in existence for ten to twenty thousand years — the time frame that would take us back to the Pleistocene — it was essential for the pond sediments to have accumulated in place, without disruption. It was those sediments that we had come to explore, to delve

3

into in a systematic way, in order to find the evidence of past landscapes that they preserved.

As the two-lane paved road we had been traveling suddenly narrowed to a single lane, I scrambled to find our location on the topographic map. Just then, Paul swerved to miss one of the numerous box turtles (*Terrepene carolina*) that were slowly crossing the lonely lane. Eventually, the paved road melted into a gravel road, then blended into a sandy two-track. Finally, at the end of the two-track, we came to a turnaround. Printed on a small sign with a wildflower insignia was "Cupola Pond, National Natural Landmark — do not pick the wildflowers." Paul parked the car under a large shade tree, and Newman, now energized, hopped out and grabbed the sediment probe and a few extra lengths of coring extension rod from the trailer. I checked my backpack for pencils and notebook. After spraying on tick repellent and wrapping rubber bands around our pant legs (to discourage the ticks and, later, the leeches), we prepared to hike the trail down to the pond.

As we stood on the ridge top above the pond, the mid-morning sun quickly baked off our insect repellent, leaving us feeling tacky and uncomfortable. Looking up at the opening in the forest canopy made by the cul-de-sac, I noted that the surrounding woods were composed largely of shortleaf pine (*Pinus echinata*) and scrubby blackjack and post oak (*Quercus marilandica* and *Quercus stellata*). Even without a breeze to rustle the oak leaves, the atmosphere cooled noticeably as we entered the shady forest and descended the slope to the pond. White oak (*Quercus alba*), several kinds of hickory (*Carya*), and flowering dogwood (*Cornus florida*) grew farther down the trail, giving way to ash (*Fraxinus*), elm (*Ulmus*), sweetgum (*Liquidambar styraciflua*), and finally tupelo gum at the wet pond margin. The tupelo gum trees were huge, at least a meter in diameter, and the tree trunks flared out into wide buttresses at their bases.

Together, Paul and Newman waded out into the tea-colored water, which was stained by humic and tannic acids leached from the accumulating leaf litter beneath the canopy of tupelo gum trees. As the two men checked the type and depth of the pond mud, I was left behind on the shore with dry feet, free to "botanize," to take field notes, and to photograph the plants I saw. Every few paces into the swamp, Paul and Newman paused to push their silvery probe into the muck.

"Black, organic-rich clay: one meter thick," Paul called out. I could see the mud welling up in a slurry around his knees. That was surely a good sign! By the time they were wet up to their waists, the sediment was more than three meters thick. Just beyond that, immersed up to their chests, the two men used the spare extension rod and kept pushing it ever deeper into the floor of the swamp, without reaching the bottom of the basin. There, just beyond the ring of tupelo gum trees, the lake sediment was at least six meters thick. With note pad in one hand and pencil in the other, I was jumping up and down with glee. This site was going to be a keeper!

There was no sign in the preliminary sediment cores that the pond had dried out at any time in the past — no rusty orange-colored sediment to indicate that it had ever been exposed to the air, which would have caused the plant fossils within it to decay. The sediment was rich in organic content and we could see seeds and other bits of plant remains with the naked eye. Because the sediment was so deep in such a small pond, it was also likely that these deposits had been accumulating, layer upon layer, for a very long time. Jubilantly, we cleaned up our equipment, dried out our pant legs, and headed back to the car for the trip to the Mark Twain National Forest superintendent's office. We were excited to tell the good news to those who had given us permission to take samples from Cupola Pond and who would now surely authorize us to carry out a full-scale sampling of its sediments. We had found our much sought after window into the past (Figure 1).

Over the next week, we collected our sediment samples and other data at Cupola Pond. Two field assistants on loan from the archaeological field school of Mark Lynott, National Park Service archaeologist, helped us with this work. Mark Lynott was collaborating with us, as well as with the Prices, on a research project funded by the National Science Foundation to study the prehistory, natural history, and historic archaeology of the Alley Spring/Doniphan area. With a crew of five, our coring session was successful.

Our coring "rig" (Figure 1) might have seemed makeshift, but it was eminently practical and it allowed us to sample the sediments of

Cupola Pond in a non-destructive manner. First, we floated an inflatable neoprene boat into open water just beyond the tupelo gum trees; then we inflated a second boat. We lashed the two boats together with lengths of rope, fitting a plywood platform over them to make a raft. The raft platform was a central four-by-eight-foot board with a hole six inches in diameter through which the coring device could be lowered or raised. Right and left sides of the platform were hinged, like a door, and window frames were attached to the central board, and they extended out like wings on either side, lashed onto the gunwales of the neoprene boats. When secured, this rig was very stable and gave a working surface that could support four or five people comfortably. During the day, we stored gear on the deck of the floating platform as well as in the wells of the boats. A third neoprene boat served as a tender to shuttle people, equipment, and sediment cores back and forth, which enabled us to leave the platform rig set up on the pond overnight. Each morning, Newman would be the first to venture out onto the raft. Newman's job was to check the boats carefully to make sure that they

Figure 1. Photograph of sediment coring rig on Cupola Pond, Ripley County, Missouri. The rig consists of a plywood platform lashed onto inflated neoprene rafts and is tied off on all four sides to tupelo gum trees that ring the margin of the pond. (Photograph by Hazel R. Delcourt)

were fully inflated, always remembering to keep uppermost in his mind the admonition printed on the glossy brochure handed out by the United States Forest Service. Their advice was, "If a cottonmouth snake drops into your boat, the first thing to remember is, don't panic!"

The process of sediment coring was straightforward. We used a hand-operated coring device, of a style initially designed by Professor Dan Livingstone of Duke University and modified for coring sediments of ponds and lakes by Professor Herb Wright of the University of Minnesota. Our "square-rod piston sampler" was constructed of a stainless steel tube, one meter long and five centimeters in diameter. The tube was sharpened at the bottom end, and it was fitted with a collar at the top end. A hole in the collar allowed a plastic-covered length of wire cable to be strung through the coring tube and attached to a piston made of rubber stoppers. The piston was fit snugly into the cutting end of the core barrel; it could be adjusted by tightening or loosening a screw onto which the rubber stoppers were threaded, thereby compressing or relaxing the fit of the piston inside the coring tube. A movable, squared-off stainless steel rod extended the full length of the tube and rested on the top of the piston.

When it was time to take a length of sediment core, we slid the core barrel down to a pre-measured depth below the sediment/water interface, just so it would come to rest at the top of undisturbed sediment at the bottom of the coring hole. The deeper we extended below the sediment surface, the more lengths of extension rod we had to attach, keeping careful measurements as we went. Once the sediment corer was in place, I measured a one-meter mark on the coring rod, grabbed the piston wire, wrapped it around a plywood plank, and braced myself. The most critical part of coring sediments by hand was making sure the piston was held in place. As the coring device was pushed into the mud, the piston created a partial vacuum that helped to guide the sediment into the core barrel rather than pushing it aside. Once the piston was secured, Paul and the others raised the squared-off coring rod its full length and turned it clockwise a quarter turn. We listened for the reassuring click that meant the square rod would hold in place and help us push the sediment coring tube down through the lake mud. At that point, the coring tube was empty and we were ready for the drive.

Four people jockeyed for position and a good grip on the coring pipe. On the count of three, they pushed downward in unison, being careful to complete the drive in one smooth motion and to push the pipe just the right distance, not to overshoot the one-meter mark. A mistake at this point could have resulted in the corer pushing aside previously undisturbed sediment, ruining it for further study.

As each core segment was raised, the lengths of coring rod that had been joined together to extend the reach of the coring device were unscrewed, removed, and laid in reverse sequence on the platform deck. With the final piece of pipe removed, the coring device came up to the surface. Paul then turned the square rod counterclockwise, slipping it back into the top of the corer's barrel. We then prepared to use the square rod to push against and thus extrude the core of mud, which generally pushed out smoothly like toothpaste squeezed out of a tube. To extrude the core, the coring device was laid on the wooden deck, and a length of aluminum foil lined with plastic wrap was prepared to receive the soft sediment. One person then rested while three extruded the soft core of pond mud and one described, labeled, and wrapped the core.

The cores of mud taken from deep beneath Cupola Pond were cold to the touch, in sharp contrast to the hot June weather in which we worked. Paul ran his finger along the sediment, feeling its texture and consistency. He paused to place a small dab of mud between his teeth, tasting it to determine how much silt or sand it contained. I recorded his description as "slightly silty clay." I compared the length of core with the color chips on my Munsell standard soil color chart, and recorded "dark gray-brown." Altogether, we extracted some twelve meters of sediment, all of which were remarkably similar in color and texture.

The sediments from Cupola Pond contained bits of plants that told a story of vegetation and climate change since the last ice age, a treasure-trove of information about changes in the landscape of the southeastern Ozark highlands over the past seventeen thousand years. In these sediments, Newman found pollen grains, spores, seeds, fruits, needles, and wood charcoal remains of plants that once grew on the watershed surrounding the pond, all that was left of the vegetation that had clothed the Ozark highlands landscape thousands of years in the past.

We live today in a world that is relatively free of glacial ice, with the exception of the Greenland and Antarctic ice sheets and mountain glaciers that exist at high elevations and high latitudes. We are used to thinking of the present day climate as the norm. Global climates, however, have been relatively warm only for the last ten thousand years, an interval of geologic time known as the Holocene Epoch. Before the Holocene, and extending back to some two million years ago, was an epoch called the Pleistocene (Figure 2). The Pleistocene Epoch was a time of ice ages, with intervals of cold global temperatures in which extensive continental ice sheets formed alternating with periods of warm interglacial climate like today's. In a typical glacial-interglacial cycle of 100,000 years duration, a glacial interval with progressively cooling global temperature lasted for about ninety thousand years and was followed by a warmer interglacial interval of ten thousand years. The Holocene is really just the most recent of some twenty interglacial intervals that have taken place since the beginning of the Pleistocene. Geologists have placed the Pleistocene and Holocene epochs into a formal unit of the geologic time scale called the Quaternary Period (Figure 2).

During the Pleistocene Epoch, the climate of the Ozark highlands was much colder than it is today. The landscape was forested with boreal conifer trees, the most abundant of which were spruce (*Picea*) and jack pine (*Pinus banksiana*). As the climate ameliorated, beginning about sixteen thousand years ago, changes began to take place in the vegetation on the watershed of Cupola Pond. During the "late-glacial" interval, a time of transition from ice-age to ice-free environments in high latitudes, communities of plants came together south of the melting ice sheets in temporary assemblages that were very different from any now existing in North America. In the Ozark highlands spruce and jack pine died back in response to climate warming. They were replaced in relay fashion by a series of deciduous tree species that were spreading northward, tracking the change in climate (Figure 3). One of the first of these immigrants was black ash (*Fraxinus nigra*), which dominated the slopes around the pond for a thousand years before it was replaced by hornbeam (*Ostrya virginiana* and *Carpinus caroliniana*), which in turn was replaced by hickory and oak. The landscape was not covered by dense forest continuously, however. Pollen

9

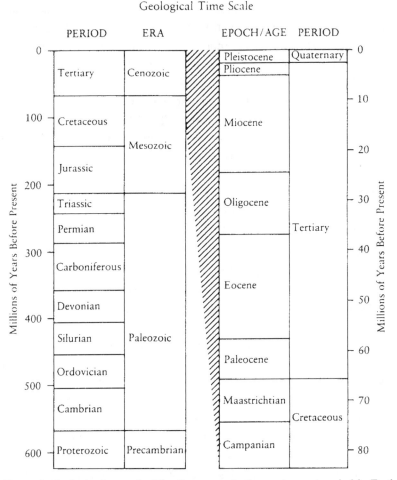

Geological Time Scale

Figure 2. Geologic time scale. The Quaternary is the most recent period in Earth history. It spans the last two million years and is divided into two epochs — the Pleistocene, from two million years ago to ten thousand years ago, and the Holocene, from ten thousand years ago to the present. (Diagram from Brouillet and Whetstone, 1993)

grains of ragweed (*Ambrosia*) became abundant about ten thousand years ago, indicating that the vegetation was then a mosaic of open groves of deciduous trees interspersed with herbs. Between nine thousand and four thousand years ago, oak trees shared space with grass (Gramineae) in a savanna-like setting. Only in the last three thousand years, for just a few generations of trees, has tupelo gum grown around

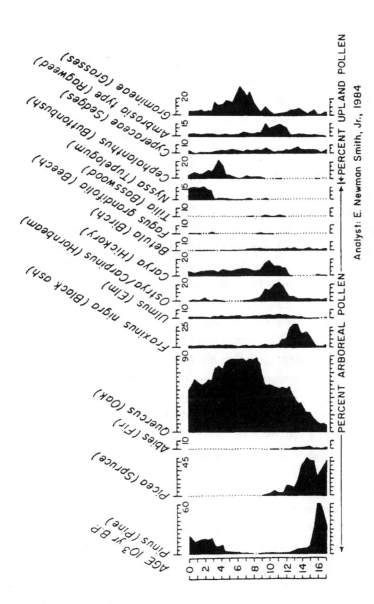

Figure 3. Pollen diagram for Cupola Pond, Missouri, indicating changes in vegetation through the last seventeen thousand years. On the diagram, the silhouettes represent changing percentages of pollen grains and spores of vascular plants as preserved in radiocarbon-dated pond sediments. (Diagram modified from Delcourt and Delcourt, 1991)

11

the shore of Cupola Pond. Similarly, shortleaf pine has become part of the landscape of the Ozark highlands of southeastern Missouri only in the last few millennia.

Reconstructing the climate and vegetation of the past requires the deciphering of clues that are sometimes quite cryptic. Learning to interpret the fossil evidence first requires not only learning to identify plant remains but also understanding the adaptations of individual plant species and plant communities to soils, climate, and other measures of environmental variation. Across eastern North America, from the boreal forest north of the Great Lakes region to the subtropical vegetation of the Gulf Coastal Plain and the Florida Peninsula, the vegetation varies greatly, but in systematic ways. The climate conditions that today control the distributions of species give indications of the nature of past climates that supported communities of plants that were far different from those seen today in a given location. We have learned at Cupola Pond and elsewhere across the southeastern United States that in order to read landscapes of the past one must first have an understanding of today's plant geography and then find the keys to decode the messages written in the fossil record.

In the remaining chapters of this book, I describe a very personal journey. This odyssey has taken me from my early education as an ecologist and plant geographer through a "time tunnel" of learning to find and interpret evidence of changes in vegetation and climate through time. Throughout my career, one thread of continuity has been my quest for understanding the history of the eastern deciduous forest and the dynamics of temperate, summer-green deciduous forest communities. Therefore, in this book, I focus on the history of the great deciduous forest of eastern North America as a case study that exemplifies the quest for understanding changes in a significant landscape system through time.

At the end of the journey described in this book, I come full circle and apply what I have learned about how species have responded to environmental changes in the past to anticipate some ecological changes that will likely occur in the future — changes that will take place in a Greenhouse world that will be much warmer than present because of human-induced, global climate warming. So this book is not just about gathering evidence with which to paint dioramas depicting life of the remote past. It is about coming to understand the dynamics of species interactions as they are forced to adapt, migrate, or go extinct during times of climate change. By finding a window into the past, we gain insights into the nature of the process by which species adjust to environmental change, as biological communities are pulled apart and reassembled in ever-changing combinations. My premise is that reading landscapes of the past gives us valuable insights to help us predict, and perhaps manage, landscapes of the future.

In Part I, I describe present-day landscape patterns across eastern North America, concentrating on the central role of the eastern deciduous forest and explaining the factors that account for its distribution and boundaries today. Then I summarize what plant geographers and paleobotanists once thought about the history of the eastern deciduous forest, before the modern era of paleoecological studies. This section of the book sets the stage for my entry into the field of Quaternary paleoecology in the context of the recent history of work in this academic field, particularly in the southeastern United States.

Part II of this book explains how I became engaged in a life-long pursuit of landscapes of the past, as a progression from interpreting present landscapes to understanding the historical factors that account for the patterns we see today. I describe the kinds of tools that plant geographers and Quaternary paleoecologists use in reconstructing changes in distributions of plants through time and in making state-of-the-art maps of changing plant communities.

In Part III, I outline what we now know about the Pleistocene refuges for deciduous forest communities, based not only upon my work and that of my husband and colleague Paul Delcourt, but also upon the studies of a number of our students and other contemporary Quaternary paleoecologists.

In Part IV, I describe the emergence of the eastern deciduous forest during the Holocene, exploring the roles of climate change and prehistoric Native American activities in shaping the postglacial vegetation. This section of the book emphasizes the dynamic nature of biological change and concludes with a summary of the lessons that Quaternary paleoecology has for projecting future consequences of continued global warming.

In the Epilogue, I discuss the potential pitfalls in trying to restore the eastern deciduous forest to a composition similar to that which existed before extensive land clearance and forest fragmentation by European Americans in the last several hundred years. In this concluding chapter, I also outline conservation measures that can be taken to lessen the negative impacts of future climate change on our increasingly imperiled eastern deciduous forest.

Part I

Appreciating Present-Day Landscape Patterns

A Whirlwind Tour of the Eastern Deciduous Forest

In the 21st century, we are privileged to be able to turn on our television sets and view Earth from the perspective of daily satellite imagery coverage. Night views capture an image of the conterminous United States bejeweled by glittering networks of cities and suburbs interconnected by the electrical grid. Interspersed among these hubs of human activity, however, are dark areas devoid of streetlights and neon signs. I am intrigued by what the photographic image can't illuminate — the fields and forests that remain mysterious in their darkness.

Numerous remnants of natural and semi-natural vegetation are interspersed within today's largely cultural landscape. Some of these remnants are located in large wilderness preserves such as Great Smoky Mountains National Park of Tennessee and North Carolina and Everglades National Park of southern Florida. Such reserves are biological "hot spots," some of the few remaining homes to many of the native plants and animals of the region.

A significant percentage of the native biota of the eastern deciduous forest region is also contained within national forest land. For example, stands of trembling aspen (*Populus tremuloides*), northern red pine (*Pinus resinosa*), and eastern white pine (*Pinus strobus*) are characteristic of Michigan's Manistee and Hiawatha national forests. To the south, a species-rich mixture of deciduous trees covers much of Kentucky's Daniel Boone National Forest, as well as the Pisgah and Nantahala national forests of western North Carolina. On the Gulf Coastal Plain, extensive plantations of both shortleaf pine and longleaf pine (*Pinus palustris*) are predominant in northern Louisiana's Kisatchie National Forest.

The remaining, and much smaller, dark areas in the satellite photos are private land holdings, including "pocket wildernesses" such as those maintained by certain paper and lumber companies, as well as lands purchased and set aside by land trusts and The Nature Conservancy. Nature preserves also surround biological research and teaching stations. For example, the Itasca Biological Station of the University of Minnesota is located on the shore of Lake Itasca, whose outlet is the rivulet that is the origin of the Mississippi River. The University of Michigan Biological Station (UMBS), located at the "tip of the mitt" in Michigan's northern Lower Peninsula, owns extensive tracts of undisturbed wetlands and a few stands of virgin pine and hardwoods. In addition, UMBS owns a large area of forest that is still growing back after the big pines, then the smaller hardwoods, were cut during the late 1800s and early 1900s. Mountain Lake Biological Station, Virginia, and Highlands Biological Station, North Carolina, are dedicated to natural history studies in the heartland of the eastern deciduous forest within the central and southern Appalachian Mountains. Sapelo Island, a barrier island located in the midst of extensive coastal marshlands of southeastern Georgia, is yet another kind of long-term ecological study site. At Sapelo Island, biologists from the University of Georgia seek to understand and to preserve the biologically rich estuarine ecosystem. At the Tall Timbers Research Station north of Tallahassee, Florida, the fire ecology of southeastern vegetation has been an area of research for more than a century. At Tall Timbers, tracts of virgin longleaf pine on sandy uplands are maintained through management with prescribed fires. At the same time, mature stands of fire-sensitive American beech (*Fagus grandifolia*) and evergreen magnolia (*Magnolia grandiflora*) trees grow to maturity on moist lowland sites that have been protected from fire. Farther south, at the southern tip of the highland spine of central Florida, and just north of the Everglades on the shore of Lake Annie, Archbold Biological Station occupies a sandhill habitat that today is rare in the highly developed subtropical Florida Peninsula.

Although satellite images and other aerial photographs might record what can be seen on the landscape, they cannot explain the relationships that exist among the landscape and the plants and animals that occur there. Nor do they convey the ways in which the landscape has been changed by people, either by those who first settled it or by those who now inhabit it. To identify and understand these relationships, we must look closer, and make use of data collected by generations of biologists, geologists, and other natural historians on wilderness areas, national forest lands, and research stations. Among the questions we might pursue, if we know where, and how, to look for answers is the story of landscapes that vanished long ago but whose legacy continues on.

Let's preface our investigation of landscapes of the past with a whirlwind tour through those of today so that we can gain an appreciation for the geographic variation among landscapes that exists at the present time. After we get a sense of how landscapes can change across space, we can turn our attention to solving the mysteries of how they have changed through time. I'll focus our attention on the present, past, and future of the temperate deciduous forest of eastern North America, a subject that has held my fascination for much of my academic career. Learning to identify and understand the variation within the landscapes of the eastern deciduous forest through both space and time has been the key to finding the locations of ice-age refuges for the individual species that now make up that forest. Finding the right methods for reconstructing and mapping past changes in tree populations has allowed Quaternary scientists to interpret the processes by which the eastern deciduous forest emerged and modern communities assembled through the past ten thousand years. Over the coming century, the continued forest fragmentation that results from ever-increasing urbanization, combined with the regional environmental changes accompanying global Greenhouse warming, portends a series of changes that will challenge the survival of temperate deciduous forest species. Through an awareness and understanding of the magnitude and direction of change that has occurred in environments and in forest communities through prehistory, we can gain a valuable perspective on the fate of these communities in the future.

Around the northern hemisphere, from North America to Europe and China, temperate deciduous forest is composed of stands of broad-leaved trees that lose their leaves and become dormant during the late fall and winter, and then break bud and leaf out once again during spring and early summer. These summer-green deciduous forests are typically multistoried. Herbs, including wildflowers that blossom in early spring before leaves are fully developed on the trees, form the ground layer. Shrubs, saplings, and small trees compose an understory beneath larger overstory trees that form an overarching canopy. Individual trees that contribute to the canopy might reach heights up to forty meters and maximum life spans of six hundred years. In PreColumbian times, before European Americans began to convert the land to agricultural fields and settlements, deciduous forest formed a nearly unbroken canopy cover that extended from the Atlantic Seaboard to the midwestern Great Plains. Deciduous forest originally covered more than 2,500,000 square kilometers (some 1,000,000 square miles), or about one-third to one-half of eastern North America.

The first map depicting the distribution of natural vegetation of the conterminous United States was published in 1858 by Joseph Henry, secretary to the Smithsonian Institution. Henry's map distinguished two types of prairie and three types of forest, and it was the first attempt to portray the geographic distribution of the eastern deciduous forest. The first quantitative map of eastern woodlands was drawn by William Brewer of Yale University for the 1870 census. Brewer's map was compiled from plat maps of the General Land Office Survey and gave estimates of forest density in acres of woodland per square mile of land. It was followed by an assessment of the distribution and value of forests, prepared by Charles Sargent for the Tenth Census in 1880 (published in 1884). Sargent's map showed the greatest density of forests in eastern North America (50 to 100 cords of wood per acre) as extending primarily through the center of the eastern deciduous forest region, from the Ozark highlands to the central and southern Appalachian Mountains, with secondary regions of high forest density located in the pinelands of the Great Lakes region and the Gulf Coastal Plain.

Plant ecologists of the early 20th century recognized that the maps prepared in the 1800s showed a fundamental relationship between climate and the distribution of vegetation. As stated by H. L. Shantz and R. Zon in their 1924 United States Department of Agriculture publication,

> *The natural vegetation is the expression of environment, it is the integration of all climatic and soil factors, past as well as present, and, therefore, if it can be distinctly and clearly indi-*

cated, provides often a better basis for a classification of environments than any one factor or set of factors...The biological unit is thus made the basis of classification and the environment is measured in terms of vegetation...

The widely cited map of vegetation compiled by Shantz and Zon recognized hardwood forest dominated by oak throughout the geographic center of the eastern deciduous forest. This map included geographic variations in forest composition across the region, mapping oak-hickory forest in the midwestern states and chestnut-chestnut oak-tulip tree (*Castanea dentata-Quercus prinus-Liriodendron tulipifera*) forest in the central and southern Appalachian Mountains. To the north, in the Great Lakes region, oak forest was described as becoming mixed with white pine and with eastern hemlock (*Tsuga canadensis*); to the south, on the Piedmont from the Carolinas to Alabama, oak shared the landscape with shortleaf pine.

The most comprehensive descriptions of virgin forest communities of the eastern deciduous forest region were those published in 1950 by Emma Lucy Braun, Professor of Botany at Ohio State University (Figure 4). Braun classified forest types and broader forest regions according to her samples of the composition of old-growth stands located on sites with rich, moist soils. Boundaries between forest types were drawn mainly at transitions in soils, landforms, and bedrock geology. She thought that the most diverse kinds of vegetation were growing in regions with the most stable geologic setting, which allowed for the evolution of plant species through long periods of time. For example, she considered that her species-rich mixed mesophytic forests, located on ancient Appalachian Mountain hillsides and within the Cumberland and Allegheny plateaus, were relict communities — ones that had originated millions of years ago in the Tertiary Period of geologic time (Figure 2) and that had persisted since their origin unaltered by changes in climate during the Quaternary Period. Here, I use Braun's description of forest regions to illustrate the kind of geographic and botanical variation that existed in the eastern deciduous forest at the time that the European settlement of North America began. Each of the nine subdivisions of the eastern deciduous forest recognized by Braun is briefly described on our whirlwind tour.

FOREST REGIONS
DECIDUOUS FOREST FORMATION

1. Mixed Mesophytic Forest Region
2. Western Mesophytic Forest Region
3. Oak-Hickory Forest Region
4. Oak-Chestnut Forest Region
5. Oak-Pine Forest Region
6. Southeastern Evergreen Forest Region
7. Beech-Maple Forest Region
8. Maple-Basswood Forest Region
9. Hemlock-White Pine-Northern Hardwoods Region

B. BOREAL OR SPRUCE-FIR FOREST FORMATION
G. GRASSLAND OR PRAIRIE FORMATION

Figure 4. Map of forest regions within the deciduous forest vegetation formation, according to the classification of E. Lucy Braun. (Diagram modified from Delcourt and Delcourt, 1979)

Centrally located and nearly covering the unglaciated Cumberland and Allegheny plateaus from southern Ohio and West Virginia south to eastern Kentucky and Tennessee, the mixed mesophytic forest region was described by Braun as the most complex plant association of the eastern deciduous forest region (Figure 4; Forest Type 1). Mature mixed mesophytic forest stands typically are dominated by a large number of canopy-forming tree species that reach heights of up to forty meters and ages over three hundred years. These species include American beech, tulip tree, white basswood (*Tilia americana* var. *heterophylla*), sugar maple (*Acer saccharum*), chestnut, yellow buckeye (*Aesculus flava*), northern red oak (*Quercus rubra* var. *rubra*), white oak, and eastern hemlock. These species-rich forests require moist, nutrient-rich

22

soils. Before European American settlement, they covered most of the land surface except for extreme environments such as dry ridge tops supporting pine and oak and floodplains of streams dissected into the plateaus that were the habitat for swamp trees such as willow (*Salix nigra*), ash, sweetgum, and sycamore (*Platanus occidentalis*).

Mixed mesophytic forest communities are multistoried, with a well-defined canopy, sub-canopy, and herbaceous layer. Lower layers beneath the forest canopy give the mixed mesophytic forest a distinctive composition and appearance. Typical understory trees include flowering dogwood, deciduous magnolias (*Magnolia tripetala, Magnolia macrophylla,* and *Magnolia fraseri*), sourwood (*Oxydendrum arboreum*), striped maple (*Acer pensylvanicum*), redbud (*Cercis canadensis*), and serviceberry (*Amelanchier arborea*). Flowering shrubs include pawpaw (*Asimina triloba*), flowering hydrangea (*Hydrangea arborescens*), and rhododendron (*Rhododendron maximum*). Spring wildflowers appearing before leafout include as many as two dozen species per square meter, with trillium (*Trillium grandiflorum* and *Trillium erectum*), dutchman's breeches (*Dicentra canadensis*), fringed phacelia (*Phacelia*), and foamflower (*Tiarella cordifolia*) forming dense colonies where locally abundant. Many species of showy ferns also line the forest floor.

Braun thought that the mixed mesophytic forest displays the greatest complexity and variation of all the forest types in eastern North America for three reasons: (1) it occupies the optimal temperature and precipitation range for the most luxuriant growth of temperate, summer-green plant species; (2) the varied geology and topography of the Cumberland and Allegheny plateaus are characterized by a wide range of aspect (direction each slope faces), elevation, and soil moisture on mountain slopes, leading to subtle differences in species composition from site to site; and (3) a great length of time has elapsed since the forests ancestral to the mixed mesophytic forest came into existence in the Tertiary Period, which has allowed for the evolution of a great variety of species. Using the general rule of thumb that the number of species sharing the canopy decreases with increasing latitude, ranging from dozens in the tropics to only a few in the northern temperate zone, Braun considered the mixed mesophytic forest as intermediate but much

23

more rich than beech-maple or maple-basswood (*Tilia americana*) forests of the upper Great Lakes region.

Mixed mesophytic forest becomes less common in all directions away from the center of its distribution in the Allegheny and Cumberland plateaus of eastern Kentucky and Tennessee as it gives way to more drought-tolerant forest to the west, to more cold-tolerant forest to the north, and to subtropical forest to the south. Between the mixed mesophytic forest region and the eastern margin of the Mississippi River floodplain is a transitional forest region that Braun called the western mesophytic forest region (Figure 4; Forest Type 2). The western mesophytic forest is less species-rich than the mixed mesophytic forest to the east, with a number of mesic trees, such as white basswood and buckeye, dropping out of the vegetation. Oak and oak-hickory forest communities replace mixed mesophytic forest communities toward the west, and glade-like openings in the forest are common, particularly on shallow limestone soils in Middle Tennessee and central Kentucky where eastern red cedar (*Juniperus virginiana*) glades and treeless "barrens" cover as much as five percent of the landscape.

West of the Mississippi River valley, oak-hickory forest extends from Canada south to eastern Texas, forming an irregular border with prairie in the eastern Great Plains (Figure 4; Forest Type 3). This forest type forms the most extensive cover in the Ozark highlands where numerous southern species of oak and hickory are predominant, some of which are not found elsewhere in the region of eastern North America. The oak-hickory forest of the Ozark highlands mantles both broad plateau tablelands and deeply dissected slopes. In southern Missouri and northern Arkansas, shortleaf pine may codominate with oaks or may locally form pure stands. Over shallow soil or bare rock outcrops, prairie openings and limestone glades interrupt the otherwise continuous cover of forest. The transition from forest to prairie at the edge of the Great Plains is characteristically an open oak woodland interspersed by prairie patches, giving way to upland prairie, with fingerlike "gallery" forests extending westward along floodplains of major river systems such as the Missouri River.

Another extensive deciduous forest type, dominated today by oak and formerly by American chestnut, extends from southeastern New

24

York along the central and southern Appalachian Mountains to northern Georgia (Figure 4; Forest Type 4). Before the twentieth century, the oak-chestnut forest region was composed of up to sixty to eighty percent American chestnut. Chestnut trees were eliminated by a fungal blight (*Endothia parasitica*) that was introduced on root stock in 1904 in New York City and spread rapidly to the south and southwest, reaching the range limits of American chestnut in Tennessee, North Carolina, and Georgia by 1938. Successional stands postdating the chestnut blight are largely composed of mixed oak and hickory and are today known as "Appalachian oak forest." In the southern Appalachian Mountains, the oak-chestnut forest type was characteristic of mountain slopes at elevations from five hundred meters to fifteen hundred meters and typically contained an understory with a well-developed heath layer including rhododendron, mountain laurel (*Kalmia latifolia*), and blueberry (*Vaccinium*). Cove hardwood communities similar in structure and composition to mixed mesophytic forest occupy the most mesic ravines within this region. Red spruce-Fraser fir (*Picea rubens-Abies fraseri*) forest grows on the ten highest peaks in the southern Appalachian Mountains, and the transition zone between oak forest and red spruce-Fraser fir forest is dominated by cool-temperate hardwood trees, including yellow birch (*Betula allegheniensis*), American beech, and sugar maple.

Braun labeled as "oak-pine forest region" a large swath of successional forest that covered much of the Piedmont from Virginia through the Carolinas and Georgia to west-central Mississippi, then extended from southern Arkansas to northeast Louisiana and into southeastern Texas between the floodplain of the Mississippi River and the oak-hickory forest region adjacent the Great Plains (Figure 4; Forest Type 5). The predominance of southern pine in forest stands composed of shortleaf pine as well as loblolly pine (*Pinus taeda*) give this forest an appearance that stands in marked contrast to that of oak-dominated forest and mixed mesophytic forest located to the north and west of the Piedmont region. The topography is flat to rolling throughout most of the oak-pine forest region. Shrubs and herbs are sparse beneath the forest canopy in comparison with forests located in the center of the eastern deciduous forest region, with blueberry, viburnum (*Viburnum*),

French mulberry (*Callicarpa americana*), and fringetree (*Chionanthus virginicus*) among the characteristic shrubs. Streams dissecting the Piedmont support a mixed bottomland hardwood forest community with warm-temperate species such as willow oak (*Quercus phellos*), water oak (*Quercus nigra*), sugarberry (*Celtis laevigata*), and yellow Carolina jasmine (*Gelsemium sempervirens*).

The southern Atlantic and Gulf coastal plains, including the extensive Mississippi embayment of the Lower Mississippi River Valley, were all considered by Braun to be part of a broad southeastern evergreen forest region (Figure 4; Forest Type 6). What the diverse forest community types of this region have in common is a preponderance of evergreen, rather than deciduous, trees. Both needleleaf and broadleaf evergreens are important. Longleaf pine forms an extensive forest of widely spaced trees on flat to rolling coastal plain uplands where wildfire is frequent. Interspersed on river bluffs and ravines protected from fire is a diverse forest of "southern mixed hardwoods," with American beech and evergreen magnolia locally abundant along with sweetgum, water oak, laurel oak (*Quercus laurifolia*), and tulip tree. In permanently wet backswamps of the Mississippi River, bald cypress (*Taxodium distichum*, a deciduous conifer) and tupelo gum dominate, with dense cane brakes (pure stands of *Arundinaria gigantea*) located on well-drained natural levees, and a mixed deciduous-evergreen forest spread across floodplain environments that are inundated by floodwaters for only part of the year.

Northern deciduous forest communities are not as rich in species as those occurring to the south. Within the beech-maple forest region (Figure 4, Forest Type 7) lying immediately to the north of the mixed mesophytic and western mesophytic forest regions and extending northward to the central Lower Peninsula of Michigan, old-growth stands are dominated by two species, American beech and sugar maple. In the northwestern part of the deciduous forest region in southeastern Minnesota, sugar maple and basswood dominate old-growth hardwood stands in the maple-basswood forest region (Figure 4, Forest Type 8). Both of these variations of deciduous forest are located north of the boundary of Pleistocene glacial ice on soils that have formed over glacial till and outwash during the past ten thousand years. Deciduous forest of the Great Lakes region is itself transitional to the eastern hem-lock-eastern white pine-northern hardwoods region (Figure 4; Forest Type 9). This forest region extends across a broad area from northern Minnesota through the upper Great Lakes and eastward across Canada and New England to New York and northern Pennsylvania and encom-passes the northern range limits of many temperate deciduous trees. Northern hardwoods forest includes sugar maple, American beech, basswood, yellow birch, American elm (*Ulmus americana*), and red maple (*Acer rubrum*), as well as aspen (*Populus*) and paper birch (*Betula papyrifera*). The northern limits of hardwood trees are determined by their tolerance to extreme low winter temperatures, with only a few being cold-hardy beyond a lower limit of -40° Celsius. The northern limit of the eastern deciduous forest, as defined by the range limits of numerous characteristic species, is located at about 48° N latitude. North of this latitude, in winter the frigid arctic air mass predominates and minimum temperatures fall below the critical threshold of tolerance for most deciduous woody plants, both for formation of antifreeze within woody tissue and for hardiness of tender leaf and flower buds.

In general, the eastern deciduous forest region is characterized by a mild to warm, temperate humid climate. Its northern and western bounds are delimited not only by changes in landforms but also by both the seasonal mean and extreme positions of frontal zones between con-trasting air masses. The interplay of three air masses is crucial: (1) the dry, frigid arctic air mass that sweeps down from northern Canada during the winter; (2) the relatively dry Pacific air mass that originates during the growing season in the North Pacific Ocean but loses much of its moisture over the Rocky Mountains before reaching the Great Plains; and (3) the moist, warm maritime air mass that emanates from the Gulf of Mexico, the Caribbean, and the western Atlantic Ocean and is responsible for most of the summer precipitation in eastern North America. Summer precipitation is most concentrated along the mari-time coastal zone and the southern Appalachian Mountains. It is also high where contrasting air masses converge over the mid-continent, along frontal zones. The positions of air mass boundaries shift season-ally across eastern North America and result in strong latitudinal gra-dients in temperature and longitudinal gradients in precipitation. Mean annual temperature and length of frost-free growing season decrease northward, and effective precipitation decreases northwestward from the southeastern United States to the central and northern Great Plains.

The southern boundary of the eastern deciduous forest may be determined by the lack of frost resistance of broad-leaved evergreen flowering trees such as the southern evergreen magnolia. Winter ice storms limit the northern extent of southern pines, which are brittle and easily toppled. Bald cypress may be limited both by the extent of suitable wetland environments and by its low tolerance of frost.

The western limit of eastern deciduous forest (and the eastern limit of prairie) is determined by a combination of increasing continentality of climate — that is, the increasing seasonal contrast in moisture and temperature that occurs as one moves from east to west in the interior of the continent, along with the influence of grazing, fire, and local soil conditions. The western limit of this forest also differs with latitude. To the north, transpiration rates of plants are lower, and sixty centimeters of annual precipitation is enough to maintain deciduous forest. Consequently, the forest extends progressively westward from Illinois to Manitoba. To the south, deciduous trees require ninety to one hundred centimeters of rain annually because of higher overall temperatures, longer growing seasons, greater evaporation of water from the soil, and higher rates of transpiration from the plants. As a result, the forest extends farther westward in humid southern and eastern Texas than in the drier interior states of Oklahoma, Missouri, and Iowa.

The nine forest regions identified by Braun within the eastern deciduous forest are bounded on her map mainly by distinctive changes in landforms, soil conditions, and hydrology. The changes in forest composition that allow such delimitation are in part also a consequence of the responses of individual tree species to gradients in climate and natural disturbance regimes. For example, Braun's mixed mesophytic forest, which is composed largely of tree species that are intolerant of extremely cold winter temperatures, prolonged drought, nutrient-poor soil, and frequent wildfire, occupies a region in which the temperature range is equable, precipitation is abundant, soils are nutrient-rich, and wildfire is rare. In contrast, on the northern and western fringes of the eastern deciduous forest region, forest types are characterized by fewer species, either those tolerant of winter extremes in temperature, high summer evapo-transpiration, nutrient-poor soil, or frequent wildfire. Southeastern evergreen forest that replaced eastern deciduous forest to the south is dominated by species that are tolerant of disturbance such as wildfire but are intolerant of frost.

The beauty and richness of the eastern deciduous forest have drawn generations of investigators to examine and explain the forest's diversity, function, history, and future. In this whirlwind tour of the eastern deciduous forest, I have emphasized a general description of the communities that today dominate each of Braun's forest regions. I have used Braun's map to illustrate the extent and general pattern of diversity among major forest types. Braun's summary is still unsurpassed in the richness of its description and its geographic coverage, but in the years since it was published it has raised more questions than it answered in respect to the origin and history of development of the forest communities of temperate eastern North America. Despite the great amount of work subsequently dedicated to cataloging the botanical diversity, ecology, and biogeography of the eastern deciduous forest, with contributions by numerous investigators who have followed in Braun's footsteps, there is still much to be learned about its natural history. Some of these questions that still begged for answers when I began my career in the 1970s focused on the origin and history of the eastern deciduous forest. As more information about the contemporary composition, environment, and ecology of this great forest accumulated, speculations about its origin and history were debated frequently and opinions were jealously guarded. To resolve the debate, it was necessary to collect data about forest history that would constitute definitive tests of alternative hypotheses about the origin, ecology, and change through time of the eastern deciduous forest. Collecting such data required developing novel methods of investigation. What follows is one career-long journey in pursuit of that history as a means to predicting future ecological changes.

What We Used to Know about Forest History

The history of the flora and fauna of eastern North America is a topic shrouded in mystery and long debated by biogeographers. What factors are responsible for the present distributions of species? How might the centers of distribution and range limits of native species have been affected by environmental changes through time? What accounts for the high richness in numbers of species packed together in the southeastern United States?

To answer these questions, at first biologists strove to document the distributions of plants and animals and registered records of their occurrence county by county on "dot maps." They compared the range maps of different species to each other and to maps for similar species compiled for the western United States, Canada, Mexico, Europe, and southeastern Asia. Long before geologists had definitive evidence of plate tectonics as a mechanism for the coming together and drifting apart of continents, botanists were aware that the plant life of the southeastern United States resembled that of eastern China. Plant geographers speculated that the shared species were once spread around the globe during an earlier era of warm climate.

In 1951, Jack Sharp, Professor of Botany at the University of Tennessee, Knoxville, suggested that the plant life of the southeastern United States had descended from an archaic world of plants — a paleoflora — that dated back as far as fifty million years ago. Professor Sharp compared what is here now to fossil plants found by E. W. Berry, a paleobotanist who worked with the United States Geological Survey in the early 1900s. Berry catalogued the kinds of fossil leaves found in over 200 localities of Cretaceous through Tertiary age scattered around the margin of the Mississippi River valley from Texas to Louisiana, Tennessee, and Mississippi. Plants growing in the southeastern United States during the Tertiary Period were part of a more

31

extensive Arcto-Tertiary Geoflora. The geoflora concept had been introduced by paleobotanists J. S. Gardner and C. Ettingshausen in their 1879 monograph detailing the Eocene flora of Britain. This concept was promoted in the mid-20th century by R. W. Chaney, who defined a Geoflora as "a group of plants which has maintained itself with only minor changes in composition for several epochs or periods of earth history, during which time its distribution has been profoundly altered although the area it has covered at any one time may not have varied greatly in size."

"Arcto" refers to the northern hemisphere, and "Tertiary" refers to the time these ancient plants were first assembled as a flora. The Arcto-Tertiary Geoflora formed extensive temperate forests encircling the globe at high latitudes across what is today a much colder arctic region. Among the plants that made up the Arcto-Tertiary Geoflora are some we would recognize today — magnolias, water lilies, and roses, as well as many of the kinds of temperate trees now found in eastern North America. The classic concept of Tertiary vegetation history depicted a northern temperate Arcto-Tertiary Geoflora and a southern, more tropical Neotropical-Tertiary Geoflora with a broad transition zone and with boundaries that through time shifted southward as Earth's climate gradually cooled.

Throughout geologic time, the earth's changing climate has been a catalyst for the evolution of plant and animal life. During the past 400 million years, land plants evolved within a global climate that was warm about eighty percent of the time. Warm climates were punctuated by intervals of colder climates, each one a relatively brief event in geologic time usually related to the shifting positions of the continents. When continents are arranged around the equator, with oceans covering the poles, the climate on land is warm; if, however, a continent such as Antarctica moves over a pole, it provides a high-latitude land mass on which precipitation accumulates not as rain, but as snow and ice. This leads to growth of polar ice caps, nature's long-term deep freezes. Thus a worldwide trend toward cooler climate may be reinforced. Such a cooling trend has characterized the earth's climate during the past two million years of the Quaternary Period.

During the Quaternary Period, Earth has experienced a rollercoaster ride of oscillations in climate. The colder phases are called glacials, during which continental ice sheets spread over much of the North American continent and mountain glaciers were more widespread. Warmer phases are called interglacials, during which the ice sheets thinned and melted back, and mountain glaciers were smaller and more localized.

Over most of each 100,000-year glacial-interglacial cycle, the earth cools gradually, culminating in widespread glaciers. Alternating with ninety thousand years of cold glacial climate are ten thousand years of warm interglacial climate. On the average, temperatures during peak interglacial times are 5° Celsius warmer than peak glacial times. Although the progression from interglacial warmth to glacial cold is gradual, the changeover from glacial to interglacial climates occurs relatively rapidly during a late-glacial phase of only a few thousand years.

Global episodes of glacial advance and retreat are recorded both in geologic deposits directly laid down by the ice sheets and in more subtle evidence found on the sea floor. Certain plankton, for example, foraminifera, living in the oceans take up oxygen and combine it with calcium to form hard shells, or tests, made of calcium carbonate. When they die, the tests of the forams settle out through the water column and are deposited as lime mud on the sea floor. There the sediments may accumulate undisturbed for up to several million years. From samples taken from sediment cores, oceanographers can identify both the species of plankton and the relative amounts of oxygen and carbon isotopes in their tests. Certain assemblages of forams are characteristic of cold surface waters; other foram communities occur in warmer waters of subtropical or tropical latitudes. Changes in composition of fossil foram assemblages through time indicate oscillations in sea surface temperature.

The waxing and waning of ice sheets on land are indicated by changes in the representation of two isotopes of oxygen in sea water. ^{16}O is a lighter atom than ^{18}O. When snow and ice are accumulating in continental and mountain glaciers, more ^{16}O is evaporated from the surface of the ocean and precipitated on land than re-enters the sea; therefore, the relative amount of ^{18}O increases in sea water and is taken up by plankton. Forams build some of this "heavy" ^{18}O into their shells, which are made of calcium carbonate ($CaCO_3$). Conversely, during times when glaciers are melting and returning water to the sea via run-off, the ocean surface waters are relatively enriched in ^{16}O and forams use more ^{16}O in building their shells. When shifts in the representation of oxygen isotopes in fossil foram shells are dated by methods such as radiocarbon dating and paleomagnetic stratigraphy, the time series of fossil data can be plotted on a graph and interpreted as changes in sea-surface temperatures.

A cosmic pace-maker of the ice ages causes rhythms in climatic oscillations, not only on the major one-hundred-thousand-year cycles but also on forty-thousand-year and twenty-one-thousand-year cycles. In the 1930s, Yugoslavian mathematician Milutin Milankovitch proposed that changes in incoming solar radiation occur in a systematic fashion because of changes in the astronomical relationship between Earth and the sun. The seasonal round of climate is influenced by the eccentricity of Earth's orbit — that is, by how elliptical Earth's path is on its annual journey around the sun. The tilt and wobble of Earth about its own axis, each of which varies systematically through time, determine which hemisphere is closest to the sun at which season. Every sixty years, the timing of winter and summer solstices and of vernal and autumnal equinoxes precesses, or advances, by one day. This may not seem like much, but over long periods of time the precession of the equinoxes causes major changes in the amount of incoming solar radiation into Earth's atmosphere, and these changes in solar inputs drive global climate change. Today Earth reaches perihelion, the point at which it is closest to the sun, during the northern-hemisphere winter. Nine thousand years ago, however, in the early Holocene, perihelion was reached during the northern-hemisphere summer. Thus, in the temperate zone of eastern North America, early Holocene climate was characterized by greater seasonal contrast, with hotter summers and colder winters than have been the rule during the late Holocene.

How would plants and animals have responded to such wide swings in climate? Their distributions today might result from past changes in climate as well as from other causes. But how can we travel back in time to reconstruct their history? In 1916, in an address to the 16th convention of Scandinavian naturalists in Oslo, Norway, Swedish geologist Lennart Von Post outlined for the first time how fossil pollen grains could be used as a record of long-term forest history. Most temperate trees produce abundant pollen, which is readily dispersed by wind, and the shape, size, and sculptural patterns of pollen grains differ from species to species. Von Post recognized that the proportions of pollen grains in sediments should reflect the composition of the forest that produced them. He reasoned that if he counted the pollen grains of each type of tree within a layer of sediment, and then compared the pollen assemblage of that layer with the pollen spectra from other layers in a vertical series of samples taken from lake or bog sediments, he could determine how vegetation changed over time in response to climate and other aspects of environment.

Von Post is widely considered the father of the field of Quaternary paleoecology — the study of the responses of plants and animals to climate and environmental change during the Pleistocene and Holocene. Principles of Quaternary paleoecology put forward by Von Post were adopted readily in both Europe and North America by plant ecologists who began to document how the distributions of plant species have changed during the thousands of years since the end of the Pleistocene.

One area that attracted particular interest among scholars studying the changes in vegetation during the late Quaternary was the species-rich forested region of eastern North America south of the southernmost limit to which the continental ice sheets advanced during the Pleistocene. That part of North America south of the confluence of the Mississippi and Ohio rivers — the southeastern United States — had never been touched directly by glacial ice. But to what extent had this region been influenced indirectly by the great ice sheets, in the form of either regional or global climate change?

By the middle of the 20th century, a controversy was brewing among paleobotanists, biogeographers, and Quaternary paleoecologists that eventually caused a rift in the scientific community. Debate centered on what evidence was acceptable as proof of the history of vegetation of the glacial boundary in eastern North America. The battle was waged in large part north of the Mason-Dixon Line, but the controversy centered on the biogeography of the southeastern United States. In one camp was the eminent plant ecologist, E. Lucy Braun; in the other camp was Professor Edward S. Deevey of Harvard University. At stake were long-held notions about the workings of the natural world.

Well-educated and strong-willed, E. Lucy Braun was one of the few women who became professional ecologists in the early to middle 20th century (Figure 5). Among her many accomplishments was her election as the first woman president of the Ecological Society of America. Together with her older sister, Annette, a professional ento-mologist, E. Lucy Braun traveled the countryside looking for examples of undisturbed, or virgin, forest. During the early part of the 20th cen-tury, the Braun sisters hiked into nearly inaccessible back country, scaled mountainsides, and ran vegetation transects up steep slopes in their quest to document the last remnants of old-growth stands of trees throughout the eastern deciduous forest. Braun's work stands as a tribute to her perseverance and keen ability to observe and record patterns in nature.

Figure 5. Emma Lucy Braun (second from left). The photograph was taken in the early 20th century on top of Natural Bridge on the Cumberland Plateau of eastern Kentucky. In one hand she carries a suitcase and over her shoulder she holds a vasculum for collecting plant specimens. (Photograph courtesy of Ronald Stuckey and the archives, Ohio State University Department of Botany)

Braun was educated in both botany and geology, within a school of thought that emphasized gradual change in the evolution of both plants and landforms. She knew the distributions of the plants of Ohio so well that she could (and did) draw the boundaries of geologic units based on the locations of plants at the limits of their natural ranges. She named her favorite forest, which blanketed the hillsides of eastern Kentucky, the mixed mesophytic forest, "mixed" for the rich mix of species and "mesophytic" for the moist soils preferred by those plants. Braun documented the mixed mesophytic forest as among the most species-rich of all the eastern deciduous forests. She envisioned her beloved mixed mesophytic forest as composed of ancient communities of plants, untouched by glaciers and unaffected by climate change through millions of years. Braun adopted the view that the Appalachian plateaus were ancient table-lands that had been worn down gradually over millions of years; these seemingly most stable of landforms were a continuously available and unchanging stage upon which mixed mesophytic forest played out an unending drama. In Braun's mind, literally, the plants had been firmly rooted in place for eons. She considered the Allegheny and Cumberland plateaus of Kentucky and Tennessee to have been the center of the evolutionary origin of the species that composed the deciduous forests of eastern North America. She also thought that the mixed mesophytic forest was descended directly from Chaney's Arcto-Tertiary Geoflora and thus was the ancestral forest from which the other modern forest types, particularly those north of the glacial boundary, were derived.

A point of view similar to that of Braun's was championed by her contemporaries, Professor Julian Steyermark of the University of Oklahoma and Professor Stanley Cain of the University of Tennessee, Knoxville. Steyermark, renowned for his treatise on *The Flora of Missouri*, cited the high species richness of the Ozark highlands, including the numerous indigenous species of oaks and hickories, as evidence that "Ozarkia" had served as both a center of origin for deciduous forest species in the Tertiary Period and as a refuge for deciduous forest species during the Pleistocene. Cain outlined in his *Principles of Plant Geography* and in journal articles his notion that the mixed mesophytic forest communities growing in the rich valleys, or "coves," of the southern Appalachian Mountains, including his beloved Great Smoky Mountains, were remnants of the Arcto-Tertiary Geoflora.

An alternative viewpoint on vegetation history grew from discoveries of Pleistocene plant fossils in the southeastern United States. In the late 1940s, technology developed during World War II, including the mass spectrometer, was making possible a revolution in science, the use of naturally occurring radioisotopes as clocks from which the absolute ages of geologic deposits could be determined. Ed Deevey (Figure 6) was just completing his graduate education, writing a long essay titled *Biogeography of the Pleistocene*, in which he summarized virtually all of what was then known about the fossil history of plants and animals on the North American continent during the Pleistocene. Deevey realized that science was on the verge of a new era and he was excited about the possibilities for advances in the field of Quaternary paleoecology. His essay became a vehicle for speculation on possibilities that many of his contemporaries were reluctant to entertain because these ideas went against the grain of conventional wisdom.

Possibly Deevey's most famous assertion in his essay on Pleistocene biogeography was that, although there was no evidence that continental glaciers had pushed farther south than the modern-day Ohio River valley, the effects of cold climate during the Pleistocene may have permeated much farther south. He suggested that there had been major shifts in the ranges of both plants and animals and profound changes in the composition and function of biological communities on a time scale of not millions, but just thousands, of years. He also suggested that certain conifers such as spruce, a type of tree today common in boreal forests of Canada, had lived as far south as Florida and Mexico during the Pleistocene. If this were the case, then temperate deciduous forest must have been forced even farther south, with its constituent species surviving through the Pleistocene in as yet unknown refuges until climates warmed to a level sufficient to release them from their bondage.

Evidence for Deevey's speculations was scanty. In part, the fossil evidence came from reports published in the 1930s and 1940s of spruce macrofossils and pollen grains from organic deposits on the coastal plain of Louisiana and Florida. In 1938, spruce cones, assumed to be

Figure 6. Edward S. Deevey at the University of Florida. (From *Ecological Society of America Bulletin* 1988: 137)

Pleistocene in age, were reported from stream terrace deposits in the Florida parishes of Louisiana by Professor Clair Brown of Louisiana State University. In 1946, J. H. Davis, Jr., found spruce pollen in peat deposits along the intracoastal waterway in the panhandle of Florida. In a third study, Professor John Potzger of Butler University reported finding spruce pollen in the deepest sediments of springs in southeastern Texas. The last bit of evidence of profound Pleistocene vegetation change in the southeastern United States was spruce pollen found by Professor Murray Buell of Rutgers University, in bottom sediments of "Carolina Bay" lakes on the coastal plain of North Carolina. From the central valley of Mexico also had come reports of spruce pollen, discovered by Katherine Clisby and her mentor, Professor Paul Sears, of Oberlin College.

E. Lucy Braun also was aware of the published reports of fossil spruce but dismissed them all, largely on the basis that they lacked (as of that time) absolute dates. She believed that the controversial spruce pollen grains may have been blown into the sites from long distances to the north, and that the Louisiana spruce cones had been rafted down the Mississippi River by glacial meltwater — thus, these fossils did not represent plants that actually grew in southerly latitudes during the Pleistocene.

The controversy articulated by Braun and Deevey over the possible influence of changing climate on the vegetation of the southeastern United States persisted through the decades of the 1950s and the 1960s. Disagreement was fueled not by the lack of evidence, but rather by the realization that to accept the evidence offered by paleoecology would be a revolution in thinking that would forever change the way plant and animal geographers would see the natural world. Both Braun and Deevey were trained in biology and geology. Each thus had a perspective that crossed the usual boundaries of disciplinary lines. Yet these two highly educated naturalists stubbornly maintained opinions that were diametrically opposed. Braun's viewpoint of nature was static. She believed that plants remained forever in their places of evolutionary origin. In contrast, Deevey allowed for the possibility that both plants and animals could be dynamic, and that species could change their locations through time by migrating under the influence of changing climate. Given enough time for generations of plants to disperse their seeds ever southward, spruce trees could eventually grow far south of their present locations during a time of cooler climate. Conversely, given warmer conditions, birds and small mammals could disperse acorns and beechnuts northward and hence expand the limits of oaks and American beech. If at a subsequent time the climate became both warmer and drier, grasses from the Great Plains might spread eastward, favored at the expense of trees that, without adequate moisture, would die back along the prairie-forest border.

Central to the issue of the effects of environmental gradients and climate change on temperate species of plants and animals was the concept of "life zones." As articulated in the 19th century by European geographers such as Alexander von Humboldt, the life zone concept was based upon the observation that distinctive assemblages of plants and animals occupy different zones of latitude from the equator to the poles. In the Americas, treeless arctic tundra north of Hudson Bay gives way to boreal spruce, pine, and fir forest in southern Canada and to deciduous forest midway through the Great Lakes region. Temperate deciduous forest changes to semi-evergreen and evergreen warm-temperate forest on the southern Atlantic and Gulf coastal plains, then to subtropical mangroves and palms at the southern tip of Florida and through the Caribbean region. Tropical rain forest is characteristic of a broad latitudinal band centering on the equator in the Amazon River basin of South America. Similar patterns can be observed across Europe and Asia. Yet another parallel can be drawn between latitudinal life zones and altitudinal ones. In ascending high mountain ranges such as the Alps, the Rockies, and even the Appalachians, life zones range from temperate at lowest elevations to arctic-alpine on mountain summits.

The distribution of life zones is determined primarily by temperature. So, if the prevailing temperature of a region either cools or warms dramatically, what effect would there be on the distribution of life zones within the region? Several possibilities include: (1) wholesale displacement with communities of plants and animals remaining intact but moving *en masse* to a new location with favorable climate; (2) extinction of species unable to migrate to a new location and unable to tolerate climate change; (3) compression of life zones into narrower belts but with plant and animal communities remaining intact; or (4) disassembly of former communities as plants and animals respond differently to climate change according to individual and different ecological tolerances.

The concept of the dynamic nature of the plant community was proposed early in the 20th century by botanist Henry A. Gleason. Gleason believed that each modern plant species was distributed slightly differently than all others, and that each species had a unique and individual set of tolerances to climate and other environmental variables. Based upon his assertion of the individualistic nature of plant species, the possibility exists that life zones could be pulled apart and reassembled in different forms by climate change. But in the mid-20th century most plant ecologists and plant geographers still preferred a more conservative view of plant communities that advocated either intact but shifting life zones or compression of life zones under the stress of climate cooling or warming. It remained for Quaternary paleoecologists to test the long-held notions of static ecology with fossil evidence.

Yet, the first maps depicting Pleistocene vegetation patterns south of the continental ice sheets in North America were based on the presumption that intact life zones had shifted southward from their modern locations. In 1958, Professor Paul Martin of the University of Arizona published a map of Pleistocene vegetation based on only eleven fossil sites and that showed a compression of life zones allowing tundra, boreal forest, and temperate forest to coexist within the southeastern United States during the Pleistocene. Another, similar map of Pleistocene vegetation was published in 1973 by Professor Donald Whitehead of Indiana University. Although the mapped patterns still appeared to show simple displacement and some compression of life zones across eastern North America, Whitehead gave the caveat that the responses of tree species may have been individualistic, and that this would have resulted in "azonal" migrations and the pulling apart of life zones. This issue was not easily resolved given the sparse amount of fossil pollen data available.

Thus, there was one central biogeographic dilemma that both fascinated and frustrated natural historians well into the second half of the 20th century. To what extent did the earth's climate and the distribution of plants and animals change through the Pleistocene and the Holocene? This quandary has inevitably spun off a whole host of related questions. For example, if climate has changed dramatically over glacial-interglacial cycles, then what were the effects on living creatures? Did they shift south and north *en masse*? Did the collage of Pleistocene landscapes resemble the life zones we recognize today? Did those life zones contract and expand like an accordion as the glaciers waxed and waned? Or did individual species respond differently from each other to a given type of change in their environment, resulting in combinations of plants and animals in the Pleistocene that were very much unlike any we can see on the landscape today? This set of questions eventually propelled both Paul and me into the fray.

If there were no modern-day counterparts for Pleistocene landscapes, then how could natural historians reconstruct the plant and animal communities of the past so that we could form a picture of those far-distant landscapes to compare with the present? And what implications would such findings have for understanding how nature functions? Are natural communities composed of tightly woven, interconnected groupings of plant and animal species that have evolved together over long periods of time, or do biological communities disassemble under the stress of climate change and reassemble with entirely new combinations of species at the end of each glacial-interglacial cycle? Or, perhaps, we wondered, does the answer lie somewhere in between, as an ecological compromise, and, if so, how could we go about searching for evidence to document the true nature of landscapes of the past in the southeastern United States?

But there was another scholarly pursuit that was important enough for us to commit our lives to. It was not the thought of persuading other biogeographers to abandon their previous ways of thinking. Instead, it was the hope that if we could understand more about how biological communities are affected by changes in climate, we could contribute to the conservation of species for the future. The prospect of rapid global warming that is anticipated in the 21st century because of human activities would make the insight we sought especially important for the conservation of biological diversity.

As we were about to enter graduate school in the early 1970s, oceanographers were probing past climate changes in deep ocean sediments and atmospheric scientists were beginning to construct sophisticated computer models by which they could predict future climate changes. Climatologists were beginning to realize that major changes in atmospheric carbon dioxide had taken place in the past and contributed to changes in climate. During cold intervals of the Pleistocene, carbon dioxide values were lower than during the warm interglacial intervals, a consequence of the reduced amount of biomass and carbon stored on land. When carbon dioxide levels in the atmosphere are relatively high, they act as a blanket of insulation, trapping heat that would otherwise be re-radiated out of the earth's atmosphere much as a greenhouse works. Since the industrial revolution, the burning of fossil fuels and the cutting of many of the world's forests have released unprecedented amounts of carbon dioxide and other so-called greenhouse pollutants into the atmosphere.

What, we began to wonder, would become of temperate deciduous forests in an even warmer Greenhouse world, one in which temperatures might be much warmer than in today's interglacial — a "super-interglacial" as projected by climatologists such as Professor John Imbrie of Brown University? How could we use lessons from the past to help see the implications for the future of those biological communities? It all came down to how people viewed the landscapes around them — either as largely static backdrops for human affairs or as a dynamic and ever-changing drama enacted not only by human players, but by plants and animals as well.

For us to begin to decipher the complex history of the mixed mesophytic forest and other communities of the eastern deciduous forest, we first had to explore a series of smaller mysteries. We began our academic journey in the Tunica Hills in southeastern Louisiana, a place that had been pivotal in the controversy between Braun and Deevey. In the Tunica Hills, we sought to resolve a number of questions that we hoped would lead us to an overall understanding of their natural history.

E. Lucy Braun had mapped the forests of the Tunica Hills as an extension of mixed mesophytic forest onto the Gulf Coastal Plain. The vegetation of the Tunica Hills was different from the rest of the mixed mesophytic forest, however, because of the presence of southern evergreen magnolias and other warm-temperate forest species. We sought to determine the composition of the forests of the Tunica Hills before they had been cut and converted to fields of upland cotton (*Gossypium hirsutum*) in the early 1800s. Also enigmatic were the present-day distributions of species of northern plants that grew intermingled with southern plants in deep, cool ravines in the Tunica Hills. The northern plants were thought by botanists to be separated from other populations of their species by considerable distances. What could their patterns of distribution tell us about their history? The final mystery of the Tunica Hills, the one to which the others led, was that of the fossil spruce remains from Pleistocene stream terrace deposits that had been reported by Clair Brown. Unraveling that mystery, as we were to find, would be a major link in a chain of understanding the natural history of the southeastern United States.

Part II

Becoming a Historian of Trees

Louisiana's Primeval Magnolia-Holly-Beech Forest

My introduction to the mystery of Louisiana's primeval forests came as a graduate student working on a Master of Science degree in botany at Louisiana State University, Baton Rouge. In January of 1972, Paul and I had packed a suitcase and two backpacks and rode a Greyhound bus from Michigan to Louisiana. The landscape we passed on our way through Ohio was full of scenery that was familiar to me, with midwestern farmsteads interspersed with fields and woodlots. After we crossed the Ohio River into Kentucky, however, I began to notice differences — little snow lay on the ground, and the grass in the meadows was green, even in midwinter.

By the time we reached Nashville, much of the landscape was forested, and from that point south through the pine forests of Mississippi, I no longer recognized by name many of the kinds of forest trees that we were seeing. When we reached southeastern Louisiana and began to traverse the river road from New Orleans to Baton Rouge, the landscape had become a mystical world of enormous flat-lying swamplands filled with evergreen magnolia, live oak (*Quercus virginiana*), and buttress-based tupelo gum and bald cypress trees, all dripping with lacy festoons of Spanish moss (*Tillandsia usneoides*).

Entering into such a new natural setting, my curiosity grew, as did my enthusiasm for my chosen area of graduate study. Paul was enrolled in geological sciences and he was as much intrigued by the landforms on the coastal plain of Louisiana as I was with the plant life. Set down into this new world, we decided to find different, yet related, thesis problems to study so that we could share every detail of our work with each other. In so doing, over the next several years we developed from a newly married couple into an academic team, with each of us learning more because of our synergistic interaction in our studies.

We lived on the floodplain of the Mississippi River, just a few blocks from the bluff edge — the university campus is built largely on a broad, high coastal terrace elevated about five meters above the floodplain. Even though the water of the Mississippi was held back by levees, I was constantly reminded of its force. In the winter and spring of 1972, record snowmelt in the north nearly caused the Mississippi to overflow its banks; a flood would have filled our ground-floor apartment.

On a Saturday night in 1972, over candlelight at a Baton Rouge coffee house, with guitar music in the air, Paul and I reached a turning point in our lives. We had been looking at a dog-eared pamphlet, a 1938 report from the Louisiana Division of Conservation, Geological Survey, whose green paper covers were fading with age and whose pages were yellowed with the decades that had elapsed since its publication. This publication was the 12th in the series of bulletins from the Louisiana Geological Survey, and it contained reports of several investigations into the Pleistocene geology and paleoecology of the Tunica Hills, an area of high bluffs, deep ravines, and meandering streams located east of the Mississippi River and north of Baton Rouge.

Bulletin 12 had been written by several LSU professors. It included two reports by Clair Brown of the LSU Department of Botany. The first contained a series of maps depicting the modern distributions of rare plants found as isolated populations in the Tunica Hills. The second report described Brown's finds of Pleistocene-age fruits, seeds, needle-like leaves, and conifer cones, paleobotanical remains that he had sieved and sorted from samples of blue-gray clay dug from streamside terrace deposits. Professor Harold N. Fisk of the LSU Department of Geosciences described the Pleistocene-age stream terraces and their geologic context as pieced together from observations made at several bluff sites exposed at low water stage along the streams. Geology Professor Henry Howe included a description of the Tertiary-age sedimentary deposits underlying the stream beds.

Bulletin 12 was filled with black and white photographs of high bluffs overlooking stream beds choked with lush vegetation, close-up shots of individual plant fossils of enigmatic species, and descriptions of now-extinct mastodons, tapirs, and horses that once roamed the Tunica Hills. All of these images evoked a sense of allure that held us spellbound, as much for what yet awaited discovery as for what was already known about the area. Here, we realized, was a study area where each of us could explore a different facet of geobotanical research, and yet combine our efforts to unravel more of the small mysteries that remained in the Tunica Hills. We knew that night that we had each found the focus of a Master's thesis.

The Tunica Hills include portions of both West Feliciana Parish and East Feliciana Parish of southeastern Louisiana, as well as adjacent Wilkinson County in southwestern Mississippi. The Tunica Hills are named after the historic Tunica Indians whose legendary golden "Tunica treasure" was sought by the Spanish conquistadors. The Tunicas were a warlike people who were eventually driven from the area and resettled in northwestern Mississippi, near Memphis, Tennessee. The Tunica Hills are an anachronism, a place seemingly uncoupled from modern space and time, suspended forever in the past. The more Paul and I learned about them, the more curious we found their story.

Paul and I had known a little about the Tunica Hills even before we left Michigan. Upon hearing that we planned to attend LSU, both Professor Ron Kapp of Alma College, with whom Paul had studied plant ecology during summer session, and Jim Kane, botanist and biology instructor at Muskegon Community College, where I had taken my first college course in biology, mentioned the Tunica Hills as a special place to explore. The area had attracted college classes on field trips from as far away as Michigan because many of the plants and animals living there are common in the north but are found nowhere else in Louisiana.

The Tunica Hills are one of the few areas within an hour's drive of Baton Rouge that is neither flat nor permanently wet. Standing on a ridge top in the Tunica Hills, you would think you were in the southern Appalachians, with ridge upon forested ridge unfolding into the distance (Figure 7). The variety of plant life is astounding, with hundreds of species of flowering plants, including a special few that are distinctly northern and decidedly out of place in the south. Not only did these northern relict plants intrigue us, but we were also fascinated by the stream deposits in which could be found fossils of now-extinct Pleistocene mammals and of spruce trees similar to those growing in Canadian forests today.

Figure 7. The Tunica Hills at Pond, Mississippi, near the Louisiana/Mississippi state line, looking north. (Photograph by Hazel R. Delcourt)

As we began to investigate the mysteries of the Tunica Hills, we found that their more recent history also added to their intrigue. My first visit to the Tunica Hills was a midwinter picnic at Oakley Plantation, now part of Audubon State Park near St. Francisville. Oakley Plantation is a strikingly picturesque antebellum home, built fully a half-century before the Civil War. The plantation home is a classic example of what cultural geographer Milton Newton of LSU called an evolved hill plantation "I" house. It is two stories high, with exterior brick chimneys and white columns supporting a two-tiered, full-length verandah furnished with rocking chairs, and set upon a brick basement. Oakley Plantation was established in 1790, before the Louisiana Purchase, and the house was built from 1808 to 1810. Today, massive live oak trees, their sweeping limbs draped with Spanish moss, dominate its grounds.

John James Audubon completed several of his famous paintings of birds, including the wild turkey (*Meleagris gallopavo*), during his stay as a tutor at Oakley Plantation from 1818 to 1821. Audubon romanticized the settings for his birds, for example painting imaginary mountains towering high above the low banks of nearby Little Bayou Sara. His paintings as rendered, therefore, cannot be viewed as a literal depiction of the early 19th century landscape of this part of Louisiana. The entries in his diary, however, contain more useful descriptions of vegetation. Especially fascinating are his accounts of strolling through forests composed of stately southern evergreen magnolias. Later travelers and naturalists such as William Dunbar of Natchez, Mississippi, described forests in which the trunks of mature southern evergreen magnolias reached over one hundred feet, or about thirty meters, before branching to their broad-leaved, evergreen crowns. These descriptions corroborate those of Audubon.

If these accounts were accurate, the original vegetation of the Tunica Hills must have been very different from today's forests. Now there are only tiny remnants of old-growth forests. Small stands of southern evergreen magnolia and American beech trees persist only in deep ravines where they have escaped timber harvest. Once we were told by a logger who was about to bulldoze part of one of our study sites that the southern evergreen magnolia trees were of so little economic value that they were to be pushed aside in order to reach the more valuable oak and walnut (*Juglans nigra*).

Louisiana foresters proudly refer to "Louisiana's third forest." The first forest was the primeval, pre-European settlement forest. This forest was cut when the land was settled by Spanish, then French, and finally English pioneers. The second forest grew back after cultivated land was abandoned after the Civil War. Today's third forest is largely composed of orderly rows of planted southern pine, and prescribed fire as well as chemical agents are used to prevent regeneration of hardwood trees in its understory.

As I discovered at Oakley Plantation, the history of the antebellum mansions of the region is intertwined with that of the forests. Understanding the cultural history of the area is essential to deciphering the patterns of vegetation. On a broad scale, those patterns can be understood by looking at maps of the overall distributions of trees at different times in the past for comparison with the present. On a finer scale, one can see the history of land use and the after-effects of land abandonment by looking at the composition of vegetation in now-abandoned cultural settings. Cultivated shrubs and vines such as roses (*Rosa*) and wisteria (*Wisteria frutescens*) still grow around the foundations of old house sites. Abandoned fields of indigo (*Indigofera tinctoria*) and upland cotton have reverted to young forest. Stagecoach and buggy trails etched into the landscape during decades of use can be distinguished only by linear or sinuous patterns of scrubby vegetation regenerating in the midst of more mature and less disturbed forest.

In Louisiana, the basic political unit is called a parish. Louisiana parishes are the size of counties elsewhere in the eastern United States. At the time of the Louisiana Purchase in 1803, the Tunica Hills were part of the territory of West Florida, which extended from just west of Tallahassee to the Mississippi River along the coast of the Gulf of Mexico. As a result, the Feliciana parishes have long been called the westernmost of the Florida parishes of Louisiana. West Feliciana Parish is located north and slightly west of Baton Rouge Parish. It is bordered on the west and south by two great meanders of the Mississippi River, on the north by the state line separating Louisiana from Mississippi, and on the east by Thompson Creek.

Several prominent streams and bayous drain West Feliciana Parish from the high land near the state line to the south and west into the Mississippi River. Tunica Bayou is the westernmost of these, while Polly Creek, Little Bayou Sara, and Big Bayou Sara are located to the east, between Tunica Bayou and Thompson Creek. Originally the largest of these streams were navigable, supporting small steamboats that supplied the antebellum estates with provisions in exchange for bales of cotton harvested from rich upland fields. Now the meandering streams are shallow and are choked with the gravel, sand, and silt that washed in from lands that were badly eroded in the 19th and 20th centuries after the upland forests were converted into cotton fields.

West Feliciana Parish was one of the richest cotton producing regions of the antebellum south. E. W. Hilgard's 1860 report on the geology and agriculture of Mississippi, written while he was State Geologist, estimated the cotton production in the Tunica Hills area at one bale of cotton per acre. The cotton gin was tested in the Tunica Hills in the late 1700s, soon after its invention by Eli Whitney. The West Feliciana Railroad, constructed in the 1830s, was the second railroad line in Louisiana, connecting Woodville, Mississippi, as well as other cotton plantations with the port at Big Bayou Sara for transport of cotton bales via steamboat down the Mississippi River. Profitable cotton farming was made possible in this hilly terrain only because of slave labor. Many of the hillsides are too steep for modern machinery to cultivate and harvest cotton bolls. The Tunica Hills cotton industry never recovered after the Civil War. During the 20th century, much of the land was converted to pastures for cattle and the production of hay. Sweet potatoes, now a major crop, are canned in St. Francisville, and remaining forest stands continue to be harvested for timber.

Topsoil eroded rapidly after plantation owners removed the original forest cover and denuded the hill slopes. Hilgard's report revealed that in 1860 exposed hill slopes were changing rapidly from rounded knolls to steep-sided hills with knife-edged crests as the loamy topsoil washed away. Streams were eroding toward their headwaters deep into the bluffs. This headward erosion created chute-like gullies and deep ravines, some of which developed at alarming rates. Eyewitness accounts from the early 1800s recount the magnitude of these landscape changes. Grand Gulf, Mississippi, located on the bluffs overlooking the Mississippi River, was so named for one of the rapidly eroding ravines, locally termed gulfs, which ripped right through the middle of town.

In the early 1970s, while we were graduate students, several LSU professors purchased a section of land along Polly Creek near the village of Tunica in the rugged heart of the Tunica Hills. Douglas Rossman, LSU zoology professor, called this parcel of semi-wilderness the Tunica Preserve. The Nature Conservancy now manages this property. The Tunica Preserve became a favorite study area to which we returned many times to collect plant specimens, measure forest composition, and marvel at fresh tracks impressed into the sandy bottom of Polly Creek by a wandering black bear (*Ursus americanus*). It was at the Tunica Preserve that I began to realize first-hand the great changes in the abundance of plants within forest communities that had occurred since the time of settlement and cultivation of the land by European settlers.

In the Tunica Preserve, Paul and I explored the ridge tops and ravines on foot, and we trekked deep within the forest to locate stands of the tallest trees. We found immense American beech trees whose diameters approached a meter across. To document the composition of the beech forest, we laid out large sample plots by tying off lengths of clothesline to four corners of a ten-meter by ten-meter square. In each sample area, we recorded the species present and the amount of area covered by each individual tree, shrub, and herb. This "quadrat sampling" proved difficult because the ravines were knife-edged and so steep that we literally had to climb the trees to move up and down within a given hundred-square-meter quadrat.

By making a series of measurements, we soon discovered something peculiar about the stand of trees. Many of the American beech trees had large trunks but low, spreading branches, as if they had grown to maturity completely exposed to open sunlight. In a well-shaded forest, saplings lose their lower branches through disuse and self-pruning. The trees then grow tall to reach the top of the canopy where they can branch out more fully to better compete for access to sunlight. The growth form of the American beech trees on the Tunica Preserve indicated that the forest there was not an example of original vegetation. Rather, the stand had re-grown after major disturbances caused by human activities. We surmised that logging at some time in the past probably had left behind open fields with only a few seedlings or saplings of American beech that eventually had grown back.

Paul and I also expected the smaller trees to be a mixture of many different kinds of hardwoods, but instead we found mostly clumps of flowering dogwood and hornbeam sprouting up from their roots. These bunch-like clusters of trees were root sprouts, growing from cut-down stumps — further evidence that the original forest cover had been removed. I saw only a few southern evergreen magnolia trees. Had the early naturalists greatly exaggerated the importance of the supposedly majestic magnolias in these forests? Or is today's forest now fundamentally different from the original, a shadow of its former splendor?

Other plant ecologists of my generation, such as Professor Paul Harcombe of Rice University and Professor Peter Marks of Cornell University, had studied southern mixed hardwood stands in the Gulf Coastal Plain and noted the discrepancy between today's forest communities and the very different descriptions of early colonial woodlands. It is possible that early explorers exaggerated the importance of southern evergreen magnolia because of the spectacular beauty of the magnolia blossoms, which are white, multiple petal flowers the size of dinner plates surrounded by large, shining evergreen leaves. Southern evergreen magnolia may never have been very important in the vegetation. Perhaps individual magnolia trees stood out from the rest of the forest, catching the eye of the casual traveler, and therefore were remarked upon more often in journals and diaries than were other, less conspicuous species.

Alternatively, magnolia trees might have formed extensive stands that were eventually cut down, as apparently were the American beech trees on the Tunica Preserve. Repeated cuttings could have eliminated magnolia seed trees, instead favoring other kinds of trees that can resprout from old roots, such as the clumps of dogwood and hornbeam that regenerate under open-grown stands of American beech.

I began to wonder how I could choose between these two contrasting notions about the nature of vegetation patterns. From existing forests, I could speculate on the causes of the present-day patterns. But to know what actually happened and how these landscapes have changed through time, I would have to travel back in time to find information that I could compare objectively with my sampled forest stands from the 20th century. Travelers' diaries can be a useful source of information. They must, however, not be taken too literally because they often present a biased point of view. How could I peel away their bias to see the forests as they really existed? How would it be possible to go back in time and measure the composition of forests on 19th century southeastern landscapes?

One source of information that has been used very successfully to determine the pre-European settlement composition of forest vegetation over much of eastern North America is the original records of the American General Land Office Survey (GLOS). Overseen by Thomas Jefferson, the GLOS was initiated in the late 1700s in order to survey the territory west of the Appalachian Mountains prior to the expansion of the American frontier beyond the original thirteen colonies. The fledgling federal government used the land survey to deed new property titles to citizens, in part to pay off a long-standing debt in the form of wages still owed to Revolutionary War soldiers. The population census and tax bases were calculated from the land ownership records.

The procedure followed in surveying the land was straightforward. Survey teams established east-west meridians and north-south base lines. They measured township grids six miles on a side using compass lines. There were thus thirty-six square-mile sections in each township. The surveyors paced the perimeter of each square-mile section, marking corners of sections and of quarter sections. In a prairie or other clearing, the survey markers were piles of stone, but in the forest the surveyors used their axes to blaze the township and range coordinates into the bark of trees. At each section corner, they found the nearest tree in each of four quadrants to the northeast, northwest, southwest, and southeast. They blazed these trees and recorded in their survey book the name of the species, distance from the corner, direction or "bearing" from true north, and diameter of each tree. The blazed trees are called bearing or witness trees, the latter because they stood witness to the survey at quarter-section corners. In some surveys, additional line trees were recorded where encountered by the surveyors as they measured out each survey line using chains and compasses. Many early surveyors also noted the lay of the land, the type of vegetation between survey points, locations of Indian villages, lakes, points of river crossings, and types of natural disturbances to the vegetation such as tornado tracks, other kinds of windfalls, and areas burned by wildfire.

If the surveys were carried out to the letter of instruction, according to the terms set forth in the surveyors' formal contracts, they became objective accounts from which a more complete general picture of original and early-settlement landscapes can be reconstructed. One method of vegetation sampling used today by plant ecologists is called the point-quarter method. At points set at regular distances along a survey line, trees are sampled in each quarter or quadrant. This method of quantifying forest composition is based upon the GLOS and has been shown to give an accurate measure of the relative abundance of trees in a stand.

One assumption inherent in using the GLOS to measure vegetation is that there was no fraud on the part of the surveyors. Yet, in the 1840s, some surveyors in the Great Lakes region falsified records, collecting their salary but in some cases never even setting foot in the township they plat-mapped. Other surveyors may have singled out the easiest species of tree to blaze, for instance the American beech, with its smooth, thin, gray bark. This sometimes resulted in record books that indicated vast tracts of American beech growing on areas that actually were mixed hardwood forest. A still more subtle form of surveyor bias occurred in cases where a surveyor could not identify all the kinds of trees. A surveyor who knew only how to recognize pines from hardwood trees would keep records that would be too generalized for researchers to use later to understand the full diversity of forests on a landscape.

Before using witness tree records as a measure of past forest composition in the Tunica Hills, I asked many questions about the surveyors. Were they employed previously for other surveys? If their names appeared on the records of many townships, they were probably honest and reliable. How many types of trees did they recognize? Did they report common tree types as well as rare ones such as swamp titi (*Cyrilla racemiflora,* a small wetland tree), lime (linden or basswood), bigleaf snowbell (*Styrax grandifolia,* a small tree with white, bell-like flowers), or sparkleberry (*Vaccinium arboreum,* a tree-sized blueberry)? Did they describe the topography and changes in vegetation along the survey lines? Finally, was there any evidence that the survey team went

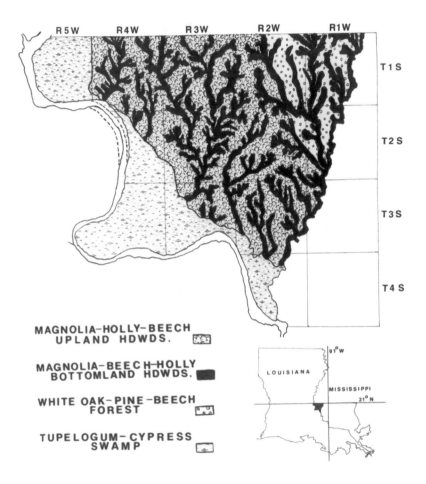

Figure 8. Pre-European settlement vegetation of the Tunica Hills, West Feliciana Parish, Louisiana. The reconstruction is based on records of the United States General Land Office Survey of 1820 to 1821. (Figure modified from Delcourt and Delcourt, 1974)

out of its way to blaze a single type of preferred tree, or were all tree species given an equal chance to be included?

In West Feliciana Parish, the American survey was accomplished in 1820 and 1821, at the same time that Audubon was in residence at Oakley Plantation. During the previous history of settlement, landowners included French, Spanish, and British settlers. Irregularly shaped sections were measured by the English metes and bounds method. In this method, boundaries of land lots were described by measuring the distance and direction to landmarks, such as prominent hills or huge trees. These land sections were established throughout the uplands and formed the basis for the layout of cotton plantations. Along the Mississippi River, long lots called arpents were measured perpendicular to the river banks, in the French style, delineating roughly parallel strips of land each of which fronted the Mississippi River and was arranged so as to include rich levee land suitable for sugar cane production. The American GLOS superimposed a township grid over this previously defined patchwork. Because the land was already parceled out and privately owned, the new survey grid consisted of only the township borders, north-south and east-west lines located at six-mile intervals, and the GLOS records give a picture of those forests that remained in early settlement times rather than those standing before the time of European settlement. The original GLOS books, containing the original field notes and the surveyors' plat maps, are archived in the Louisiana Land Office in Baton Rouge.

Reading the original survey notes revealed much to me about the day-to-day life of the men in the field. In the early days, the Tunica Hills were also known as the cane hills because of the dense growth of American cane, a New World relative of the oriental bamboo. Like bamboo, American cane spreads by underground stems, called rhizomes, to form extensive thickets of densely packed, inch-thick woody stems that in the aggregate are nearly impenetrable when mature. The first land to be cleared for sugar cane cultivation in Louisiana was the fertile first bottom, levee crests and backswamps that originally were thickly covered with these cane brakes. It was the surveyors' task to run survey lines through forests containing an undergrowth of dense cane. In the cane hills region of the Feliciana parishes, their solution to the problem was to set fire to a swath of cane along the survey transect, clearing both a pathway and a line of sight for establishing survey lines.

The surveyors evidently preferred not to work in the backswamp of the Mississippi River floodplain. Without exception, their survey posts extended into the muddy backswamp beyond the edge of the river bluff for only one or two survey locations. Apparently, dry feet were worth more to them than the money earned by challenging the swamp-lands. They encountered other kinds of adversity as well. One hand-written page in the survey book for West Feliciana Parish, dated December 1820, read faintly, "Too cold to work. Ink frozen in pen."

I used nearly 400 witness trees that were recorded in the GLOS throughout West Feliciana Parish to map the original forest types as they existed in 1820 (Figure 8). Bald cypress and tupelo gum were the most important trees in the backswamp of the Mississippi River. On both the hilly uplands and in the bottomlands of the tributaries to the Mississippi River, mixed hardwood forests predominated. The mixed hardwoods were composed mostly of southern evergreen magnolia, American holly (*Ilex opaca*), and American beech.

What I discovered after extensive analysis of the GLOS witness trees was that the early naturalists were right! Southern evergreen magnolia must have formed truly magnificent forests at the time Audubon, Dunbar, and other early residents of the Tunica Hills viewed them. Those forests flourished on the most fertile soils, which were confined to a relatively narrow zone along the eastern bluff of the Mississippi River. Moist, nutrient-rich, loamy silts formed not only the soil but also the banks of streams and the vertical walls of deeply entrenched roadways such as Tunica Trace in the Tunica Preserve. The silt-loam substrate is tens of meters thick at the edge of the Mississippi River bluffs, but it thins to less than one meter thick in the eastern part of West Feliciana Parish. There, the pre-European settlement forests were composed of oak and longleaf pine, trees that are tolerant of dry, nutrient-poor soil and can withstand occasional wildfire.

In 1850, the 31st parallel of north latitude that marks the state boundary between southeastern Louisiana and southwestern Mississippi was resurveyed. Many of the new corner posts were set in cotton fields or on plantation grounds. Ornamental shade trees, such as walnut and white mulberry (*Morus alba*, which may have been planted as a trial grove for raising silkworms) were common and served as bearing trees. In only thirty years, the once widespread tracts of primeval American beech, southern evergreen magnolia, and American holly forests had been sacrificed to king cotton. This forest has not since been restored to its original magnificence.

The complex history of the Tunica Hills forests cannot be understood just by looking at the modern landscape. As I delved deeper into the mystery of the apparent paradox between the forests early settlers saw and those I encountered in the 1970s, I began to learn how limited my first impressions had been. In the process of converting the Tunica Hills from a natural landscape to a cultural one, not only was the magnolia-holly-beech forest eliminated, but also a cycle of erosion was begun that transformed the hill slopes from gently rolling land to knife-edged ridge tops and steeply dissected ravines. Streams used in the early 1800s for shipping cotton and other goods became no longer navigable by shallow-draft steamships as eroding topsoil was turned into stream sediment. After the Civil War, once-magnificent southern evergreen magnolia forests were scarce. The properties of stately mansions became overgrown with second-growth shrubs and trees which left plantations such as Rosedown, Oak Alley, and Oakley defined only by remnants of gardens and by shady lanes lined by mature live oak trees. One legacy of king cotton in the Tunica Hills is a landscape forever changed.

Searching for Spurges in the Tunica Hills

East of the wide valley of the Mississippi River, hills capped by silt-loam soil line the edge of the river bluffs. Silt forms the bluffs — silt that westerly winds swept up from the Mississippi River valley and deposited to the east in thick layers during the Pleistocene. This highly erodable, wind-blown silt is called loess. In 1846, the eminent British geologist Sir Charles Lyell described in his diary what he had seen while touring on a riverboat from Baton Rouge past St. Francisville and up to Natchez, Mississippi, during his second trip to the United States:

> At Natches, there is a fine range of bluffs, several miles long, and more than 200 feet in perpendicular height, the base of which is washed by the river. The lowest strata, laid open to view, consist of sand and gravel, destitute of organic remains, except some wood and silicified corals and other fossils, which have been derived from older rocks; while the upper sixty feet are composed of yellow loam, presenting as it wastes away, a vertical face toward the river. From the surface of this clayey precipice are seen, projecting in relief, the whitened and perfect shells of land snails...the resemblance of this loam to the fluviatile silt of the valley of the Rhine...which is generally called 'loess' or 'lehm'...is most perfect.

Lyell's was the first recorded observation of such loess soil in North America.

Today, Europeans spell the word loess by putting an umlaut over the letter "o" to read "löess." When Midwesterners pronounce this word it sounds like "luss." Texans say the word as if it is spelled "lurse." In Louisiana, geologists call it "low-ess," with the emphasis on the second syllable. The Tunica Hills are sometimes called the loess hills. Near Natchez, Mississippi, the bluffs are known as the walnut hills

65

Figure 9. Hazel Delcourt at the Tunica Trace, the old road bed through the western part of the Tunica Hills. (Photograph by Paul A. Delcourt)

because of their verdant forest growth. In the 1800s, wagons traveling along trails between Natchez and Baton Rouge wore ruts into the loess. With continued use, the roadbeds were gradually worn down into the hillsides. Eventually this erosion left steep loess bluffs on either side of the road, which was then called a trace. The most famous of these old roads is the Natchez trace, which today is a National Parkway extending from Nashville, Tennessee, through much of the state of Mississippi to Natchez.

The most spectacular old roadbed in the Tunica Hills is called the Tunica trace (Figure 9). This winding, single-lane track connects two otherwise isolated hamlets. Old, gnarled "haricane" trees overarch the trace like a tunnel. Tunica trace is flanked by orange-brown cliffs that rise as much as five meters above the roadbed. Just as Lyell observed long ago, ghost-like fossil shells of land snails, entombed within their silty crypt, still gleam white where the high walls are eroded to expose their nugget-like forms. With enough time, rainwater can gradually seep through the soil layers, dissolving away small chips of limestone rock. Drawn downward around tree roots, the carbonate once again

66

precipitates in multiple layered tubes and irregular nodules of lime. The clumps of lime bind together to form hand-sized figurines — loess "dolls" that have been played with by generations of Tunica Hills children.

Loess is a remarkable substance. Keep it dry, and it is very strong and can maintain a perfectly vertical face, making for a sheer cliff that is almost impossible to scale. But get it wet, and it turns into a slurry of mud and water that begins to flow like an amoeba, oozing down the hill slope. These mudflows are called solifluction lobes. Once rain-saturated, solifluction lobes flow down at intervals into Tunica trace, burying the road and making it impassable until someone comes to plow it out.

We had several close encounters with flowing loess. One winter Doug Rossman decided to put a mobile home on what he thought was a stable, flat geologic bench perched at the edge of a remote, deeply eroded ravine in the middle of the Tunica Preserve. He planned to use the trailer as a field station for his observations of wildlife on the nature preserve. One day, after a winter storm that brought heavy rain, the trailer began to glide slowly downhill. Also pulled down the slope were several large tulip trees. When we arrived to help rescue the trailer, tree roots were snapping and the entire hillside appeared to be flowing out from under us. We cut saplings and used them as rollers to slowly push the trailer away from the ravine as the bank of wet loess flowed out from under it.

Ravines deeply eroded into the loess hills provide an ideal habitat for a number of rare vascular plants. One of these unusual plants is called Allegheny spurge (*Pachysandra procumbens*). This spurge is a sprawling plant with round, toothed leaves the size of a human hand (Figure 10). The leaves are mottled green and white, and they have the feel of living velvet. In Tennessee, Arkansas, Mississippi, Alabama, and Louisiana, it grows in clones up to several meters across. Colonies of Allegheny spurge show up only in isolated places, always along stream banks. The one species of Allegheny spurge living in the southeastern United States has several close relatives growing in southeastern Asia. *Pachysandra terminalis*, a Japanese species, is grown in commercial nurseries in the United States and is used commonly as a ground-

Figure 10. Allegheny spurge (*Pachysandra procumbens*), a rare plant found in the understory of mixed mesophytic forest, in the Tunica Preserve, West Feliciana Parish, Louisiana. (Photograph by Hazel R. Delcourt)

cover plant in landscaping. The extreme east-west disjunction in the native ranges of these similar biological species in the genus *Pachysandra* most likely resulted from the separation of populations of an ancestral species dating back millions of years to the Tertiary Period. Then spurges were part of the Arcto-Tertiary geoflora and were probably widely distributed throughout the northern hemisphere.

In the Tunica Hills, Allegheny spurge plants live along the banks of Polly Creek and in dissected terrain at the end of a series of anastomosing loess ridges in the heart of Tunica Preserve. In the spurges' favored habitat, tangles of vines line deep, wet, cool, shaded ravines filled with the earthy odor of fresh leaf litter. There also is found a lush growth of ferns, including the rare northern maidenhair fern (*Adiantum pedatum*). The habitat is an intricate living maze, scaled by climbing tree limbs and grabbing onto interwoven rooted handholds. Beneath the overstory of American beech trees are smaller trees of flowering dogwood and hornbeam, with occasional sugar maples, deciduous magnolias, and southern shrubs with colorful names such as French mulberry, sparkleberry, star anise (*Illicium floridanum*), silverbell

68

(*Halesia diptera*), and bigleaf snowbell. The ground cover is reminiscent of a northern beech woods — spring wildflowers such as wild trillium and Carolina spring beauty (*Claytonia caroliniana*) abound there but are otherwise rare in southern Louisiana.

The deep ravines contain other remarkable plants besides the Allegheny spurge and the northern maidenhair fern. Also found there is the fanciful, white-berried doll's eyes (*Actaea pachypoda*). Ginseng (*Panax quinquefolium*), a medicinal plant that has been collected nearly to extinction, grows there, as does the bittersweet vine (*Celastrus scandens*) and the very rare southeastern United States endemic plant, star vine (*Schisandra glabra*). Most of these plants have only one species growing native to North America today, and all share an unusual geographic distribution in the southeastern United States. Because the Allegheny spurge occurs in isolated patches and in only a few locations in the state of Louisiana, botanists such as Clair Brown thought that the geographic distribution of this plant is evidence that the Tunica Hills were an island-like biological refuge. For some unknown reason, these special plants have been able to survive there through a long interval of earth history. Perhaps, it was thought, millions of years ago the Tunica Hills were a high piece of land surrounded by ancient Tertiary seas. The loess hills have thus long posed an enigma, as they are a separate and little understood place.

E. Lucy Braun featured the story of the Tunica Hills plants in her 1950 book, *Deciduous Forests of Eastern North America*. To tell the story of these plants, Braun relied on the dot-maps of their distributions that had been prepared by Clair Brown and published in *Bulletin 12*. Those maps showed that a suite of northern plants appeared to be separated from their main ranges to the north by distances far greater than their natural dispersal ranges. Braun accepted the modern distributions of the northern disjunct plants as proof of the antiquity and relative isolation of the Tunica Hills landscape. She believed that the Tunica Hills flora and vegetation had remained unchanged since the end of the Tertiary Period some two million years ago.

But were these plants really so isolated? If so, it would imply a long history of geographic separation from other populations of the same species. Or can their present-day distributions be explained more simply, as a result of incomplete exploration by botanists? For me, the mystery of the disjunct distribution of the northern plant species in the Tunica Hills became the object of a fascinating geobotanical treasure hunt.

At first it seemed unlikely that botanists would have missed much. Since the 1700s, the southeastern United States has been subjected to intensive botanical exploration. Among the famous early plant collectors was the father and son team of John and William Bartram. They searched for such rare plants as Oconee bells (*Shortia galacifolia*), with its delicate, white, bell-shaped flowers, native to only a few mountainous locales in northern Georgia and South Carolina. The Bartrams were among the few botanists to see the only known grove of Franklin tree (*Franklinia altamaha*) in coastal Georgia before that species became extinct in the wild. For hundreds of years, botanists have scoured the southeastern United States looking for such oddities of the plant world — the region as a whole is world-famous for its rich and diverse flora. Given this history of exhaustive botanical investigation, how could earlier botanists like the Bartrams and Clair Brown have missed finding these out-of-place northern plants if they did grow between Louisiana and their main ranges in Tennessee and northward?

Finding the answer to this question required developing a geobotanical strategy that would include an intensive search for spurges and their associates. Essential to this quest was finding living populations of the plants in enough locations between the Tunica Hills and Memphis, Tennessee, that we could be certain the species were geographically and reproductively linked.

Paul and I were very familiar with the habitats of the plants in the Tunica Hills. There they are isolated and difficult to find because they grow in deep, protected ravines eroded in the rich silt loam of the loess bluffs where the casual traveler would never see them. Paul surmised that the loess is a key to their persistence, both because it is fertile and because it erodes into vertical ravines, creating a moist and cool environment. This distinctive environment is more common in Tennessee than it is in Louisiana, which is generally characterized either by dry sandy coastal-plain uplands or by wet swamplands.

We traced the loess deposits on geologic maps. The windblown silt drapes the edge of the Mississippi River bluffs with loess deposits tens of meters thick. It is often exposed as a sheer cliff face along the crest of the bluffs. Within a few tens of kilometers to the east, the loess thins sharply, to depths of a meter or less. Beneath the loess blanket in the Tunica Hills lie deposits of sand and gravel, remnants of Tertiary and Quaternary coastal terraces. Older geologic deposits underlie the loess to the north along the rest of the margin of the Cretaceous Mississippi Embayment. Erosion and dissection of the land surface is greatest in the thickest loess, near the bluff edge.

We outlined the bluff edge on topographic maps that covered the area from southeastern Louisiana through Mississippi to Tennessee. We identified prospective sites in which Brown's "northern" plants might be found in the region of thick loess by noting locations where the steepest slopes were covered with natural vegetation. From the contour maps, we found that, topographically at least, the Tunica Hills are not an isolated island. Rather, they form the southern terminus of a continuous belt of hilly land blanketed by loess. In fact, the loess hills extend continuously from north to south through a significant portion of the states of Illinois, Kentucky, Tennessee, Mississippi, and Louisiana. We began to call the whole region of loess hills "the Blufflands."

If the Bluffland habitats are continuous, we reasoned, then so might be the natural distributions of these so-called out-of-place northern plants. What remained to be discovered was whether the assemblage of plants grew in enough patches of the right habitat to connect their distributions as continuous from north to south along the Blufflands.

During the spring break of 1973, we packed our maps and plant presses and headed north on I-55 toward Memphis. We were bent on solving the mystery of the spurges as well as those of the other "northern" plant species, by documenting their occurrences in the Blufflands between Tennessee and Louisiana. Our topographic maps showed several areas of greenery immediately to the south of Memphis, and we headed for the nearest ravines, expecting to fill our collecting bags with spurges and other northern plants. But soon it became apparent that our maps were out of date. One by one, we crossed a number of prospective plant-collecting areas off our list. Suburbs of Memphis sprawled out onto the hillsides. Nearly all the deepest ravines in the most rugged topography were drowned behind earthen dams constructed to form ponds and small recreational lakes.

From Memphis, we traveled US 61 south along the edge of the Mississippi River bluffs. US 61 is the old road that has long connected the port of Memphis with cotton plantations to the south. Once we started our work in Mississippi, Paul and I soon realized why botanical exploration in the state had not been as complete as elsewhere in the southeastern United States. It had to do with land ownership laws and legal rights against trespassing. We always tried to ask permission before crossing anyone's land. But in Mississippi, gaining permission to collect wild plants presented a challenge. Any citizen had the legal right to boat or swim or walk in any navigable stream — that was public property. The loess-lined ravines beyond, however, were often posted, as was much of the privately owned land in Mississippi. We found it frustrating that in many areas of the Blufflands, individual parcels of forested land were often separated from the landowner's residence, making it impossible for us to check each and every place where our plants might have been tucked away.

We did, however, manage to make some collections in each county on our list, and we found that not only Allegheny spurge, but also northern maidenhair fern, white-berried doll's eyes, bittersweet, and even the more elusive ginseng and star vine generally grew together in the Blufflands loess habitats throughout the western part of the state of Mississippi. At each collecting site, for each specimen we collected, we recorded the scientific name of the plant, the details of its location, and the identities of other plants with which our specimen was associated. First we placed a whole plant, roots and all, between sheets of newspaper, then surrounded it with thick paper blotters, and finally bound it between wooden covers cinched up with belt straps. We took all of the drying specimens back to LSU in our plant presses. When the specimens were fully dried and fumigated to protect them from mold and insect damage, I mounted them specimens on acid-free sheets of oversized herbarium paper, labeled them, and added them to the LSU herbarium as vouchers to document our collections. These herbarium specimens are archived as permanent records.

During our ten-day collecting trip, we explored at least two or

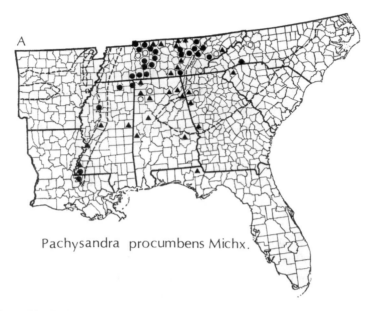

Figure 11. Distribution map for Allegheny spurge (*Pachysandra procumbens*) show-ing counties in which it has been collected in the southern Appalachian Mountains, the Blufflands, and elsewhere in the southeastern United States. (Diagram from Delcourt and Delcourt, 1975)

three promising sites in each of the counties bordering the eastern side of the Mississippi River valley. Our geobotanical strategy in the Blufflands worked well enough to establish that the rare plants do have more or less continuous distributions from north to south (Figure 11). Our successful search for spurges and their associates reinforced the importance of the Tunica Hills and the Blufflands as sanctuaries for many species of plants. But rather than an isolated island refuge some-how cast adrift unto itself for millions of years, we began to view the Tunica Hills as more like the southernmost extension of an archipelago-like scattering of habitats that link the populations of these plants to-gether along the Blufflands.

What we still didn't know was how the history of development of the loess hill habitat was related to the history of changes in the distri-butions of the plants. Following the Tertiary Period, two million years have elapsed during which dramatic changes in landforms, soils, and biota have taken place around the world. In central Europe, during the

Pleistocene, wind-blown loess was deposited uniformly across a nearly treeless steppe plain, an open landscape across which the silt was free to blow without a baffle of tall plant growth. In the Blufflands of the eastern United States, however, the loess is draped symmetrically over the hills. It is thickest on top as well as near the bluff edge adjacent to the Mississippi River floodplain, and it thins dramatically to the east. Was this distribution of loess evidence that Pleistocene vegetation in the Blufflands was a forest that acted as a baffle to entrap the wind-blown silt? Was the climate significantly colder during the Pleistocene, allowing northern species to spread southward? Did they become stranded in the Tunica Hills and the Blufflands when Holocene inter-glacial warmth spread across the southeastern United States? Were only certain cold-adapted species affected by the climate warming of the past 10,000 years in such a way that they became extinct in the southern part of their ranges, while other, more temperate species such as the Allegheny spurge persisted to the present day in the cool, moist Blufflands ravines?

Finding the answers to these questions meant becoming more than a short-term historian of trees. It meant tackling head-on the longstanding controversy in opinions between biogeographers and Quaternary pa-leoecologists. Even in the early 1970s, most plant geographers remained convinced that changes in the distributions of plants required lengthy adaptation and adjustment to gradual geological processes and long-term global changes in climate and plate tectonics. Quaternary paleo-ecologists, on the other hand, were more open to the possibility that rapid and dynamic changes in the biota can take place over shorter periods of time. Thus, having become used to thinking in tree time, we began the next stage of our quest to understand the long-term history of the biota of the Tunica Hills and the Blufflands.

Hunting Mastodons in
Louisiana Bayous

In 1973, a conference was held at LSU on the ecology of the Pleistocene. At this symposium, we heard lectures by several notable Quaternary paleoecologists. Each lecturer held the audience spellbound as stories unfolded of dynamic interchanges of plant and animal species as they adjusted, moved about, or became extinct during past episodes of rapidly changing climate.

Paul Martin spoke on the possible causes of large-mammal extinction. He outlined the evidence for rapid environmental change at the end of the Pleistocene as possibly causing widespread disruption in faunal communities, and he argued for an alternative explanation involving the "overkill" of large Pleistocene mammals by Paleoindian hunters who invaded the Americas from Siberia some 12,000 years ago.

Professor Margaret Davis, then of Yale University, presented maps of tree migrations over the past 20,000 years. She emphasized the apparent individuality in the rates and directions of spread of boreal and temperate trees as they followed the retreating margin of glacial ice. She called for increased information on the Pleistocene distributions of tree species in the southeastern United States, for which the ice-age refuge areas were as yet largely uncharted.

Professor Herb Wright of the University of Minnesota gave his views on current problems and future directions for research in Pleistocene ecology. Professor Wright reminded us all that any conclusions about changes in plant and animal communities must be tempered by the realization that

> . . . the vegetation of the south-central United States during [the late Pleistocene] — south of Illinois between the Appalachians and the Great Plains — is virtually unknown. The vast deciduous forests that prevail in the Middle West and the East today may have been restricted to local areas during the glacial period. . . .

The words of professors Martin, Davis, and Wright inspired us, and the mystery of the history of deciduous forests in the southeastern United States became one of the many puzzles Paul and I dedicated ourselves to solving.

In a way, the words of the seminar speakers gave us license to hunt mastodons. From the accounts published in *Bulletin 12*, we knew that nearly complete mastodon (*Mammut americanum*) skeletons had been recovered from Tunica Bayou and other stream beds in West Feliciana Parish. Those skeletal remains, residing in glass cases lining the entrance of the LSU Geosciences building, attested to the former presence of these behemoths in the Tunica Hills. Great proboscideans, now extinct, lived in southern Louisiana more than 10,000 years ago during the late Pleistocene. We, along with our symposium speakers, wanted to learn what Louisiana was like when mastodons lived and glaciers covered the northern half of North America. No doubt Louisiana's climate and vegetation were very different then. But how different were they from present conditions? The vision of elephant-like beasts living where swamps of tupelo gum and bald cypress now grow seemed incongruous. Across the Great Lakes region from Wisconsin to Michigan and Ontario, most mastodon remains were found with fossils of spruce and tamarack (*Larix laricina*) trees. Certainly, we all believed, mastodons could not have lived together with Spanish moss and southern evergreen magnolia.

Most — but not all — of the plant fossils originally reported in *Bulletin 12* were identified as species that grow in the Tunica Hills today. Certain fossil cones reported in the bulletin were similar to those of white spruce (*Picea glauca*), a tree which today grows much farther to the north, in the northern Great Lakes region and in Canada. During the Pleistocene, then, our symposium guests inquired, was there a strange mixture of boreal species intermingling with temperate plants? If so, how did the climate change? What could account for both northern and southern species growing together in assemblages quite unlike any found on modern landscapes? In particular, could the fossil remains of white spruce have been merely rafted down the Mississippi River from far to the north and washed up into the mouths of streams feeding into the

Mississippi River from the Tunica Hills? If this were the case, the fossil assemblage would not represent a biological community in which all the species lived together at the same place and at the same time. Instead, the fossils would include plants and animals from two completely different landscapes separated in time and/or space but whose fossils were intermixed where they were buried in the clays of the stream bed. In this eventuality, they would not be evidence for climate change at all.

One of the most important tasks in unraveling the mysteries of Pleistocene time in the Tunica Hills was to verify the original identifications of the plant fossils and to replicate the important finds of Clair Brown. First we needed to learn to recognize fruits and seeds, needles, cones, and leaves of both northern and southern species of plants that we were most likely to find in fossil form. For this task, we used the species list and photographs in *Bulletin 12* as a guide. We also interviewed Brown, who in 1973 was an emeritus professor at LSU. Professor Brown told us stories of his collecting trips to the Tunica Hills with geologist Doc Howe, and about their visits to remote places with names like Weyanoke and Pinckneyville. He related his experiences in discovering the unexpected evidence of northern and southern fossil plants. Then he awed us by showing us one of the dried spruce cones that he had long ago photographed for *Bulletin 12*.

Our next step was to go out into the field, to revisit sites such as Percy Bluff on Little Bayou Sara where the spruce cones had been found nearly forty years earlier (Figure 12). The cut bank from which the spruce cones had come also contained mastodon tusks. Thus, the Percy Bluff locality was crucial to understanding whether mastodons, white spruce trees, and warm-temperate plants had co-existed in the late Pleistocene. Understanding whether boreal fauna and flora were contemporaneous with a temperate biota was essential to determining the nature of climate change in southern Louisiana and to resolving a long-standing scientific debate.

But first we had to overcome certain logistical problems in gaining access to the Percy Bluff site. There was no bridge over Little Bayou Sara close enough to make the Percy Bluff site easily accessible on foot from a public crossing. So on the hot Louisiana morning when

Figure 12. Clair Brown at the original Percy Bluff fossil locality, a stream-cut bank on Little Bayou Sara, West Feliciana Parish, Louisiana. (Photograph reproduced from Fisk *et al.*, 1938)

we began our mastodon hunt, we parked our Toyota Land Cruiser near a bridge crossing considerably downstream from where Brown had marked the site of Percy Bluff on our map. We prepared our fossil-collecting gear of shovels, backpacks, canteens, and camera, and planned to wade in the stream bed to either the Percy Bluff site or a cut bank as close to the original site as possible. Once at our site, we would excavate what we hoped would be fossil-laden Pleistocene clay deposits.

Where we entered the meandering Little Bayou Sara, the stream banks were near-vertical cliffs of greenish clay. The stream bed contained a thick layer of gravel. After we had walked bout half a kilometer in the direction of the headwaters of Little Bayou Sara, the stream bottom changed character. Gaps appeared in the loose gravel where the waters of Little Bayou Sara had scoured out pools, and we walked on a layer of the firm, green clay. We followed this clay layer around the next bend. Paul checked the map and confirmed that we were walking on ancient marine mud, what Fisk and Howe described in *Bulletin 12* as "unfossiliferous green Miocene clay of the Pascagoula Formation."

Farther upstream, the appearance of sand and gravel in the cliffs marked the transition, or contact, between the older, deeper Miocene marine clay and the younger, higher, terrestrial deposits of a coastal terrace of late Pliocene to early Pleistocene age. The orange, red, and purplish colored sand and gravel were characteristic of the Citronelle Formation, which is distributed widely across the Gulf Coastal Plain from southeastern Texas to southern Georgia. Around several more meander bends in the stream, we saw a change in the color of the sand and gravel deposits to yellows and tans, which were overlain by a thick layer of buff-colored silt loam. This change in the composition of the stream banks marked a place where Little Bayou Sara once had meandered and cut into the Citronelle deposits, then filled in the meander cut with sediment that we guessed would be late Pleistocene in age.

Based on the changes in geologic setting and our location as paced out following the topographic map, we deduced that we were close to the original location from which was taken the classic photograph of Pleistocene plant beds and mastodon tusks weathering out of Percy Bluff (Figure 12). Using *Bulletin 12* as our field guide, we continued to search for fossil-bearing Pleistocene clay. By previous accounts, this clay should be blue-gray in color and it should lie at the base of the stream bank.

We knew that blue-gray clay layers were exposed at the water line not only along Little Bayou Sara but also along Pinckneyville Branch in the headwaters of Little Bayou Sara, and along Tunica Bayou. These clays previously had proved to contain a rich assortment of mammal bones such as peccary (*Platygonus compressus*) jaws, horse (*Equus*) teeth, and mastodon tusks and femurs. The fossil beds also contained wood, twigs, leaves, nuts, and other remains of plant life that had lived there and were buried more than ten thousand years ago. We found that the clay was tinted blue only when freshly exposed to the air. It then quickly oxidized to a dull gray color, appearing much like ordinary pottery clay.

Along the streams in the Tunica Hills, only the layers of blue-gray clay, shaped like the lens of an eyeglass, thin at the edges and thicker toward the middle, preserve fossil leaf mats. The clay lenses represent what once were quiet backwater pools in which fine-grained sediment settled and into which leaf litter drifted — litter that otherwise would have rotted away on some ancient forest floor. Along the streamside, however, the leaf mats became compacted, then they were buried by overwash of the ancient Little Bayou Sara as it meandered across its floodplain, obscuring some of its former deposits even as it eroded and exposed others.

After wading around several more meanders, we located a cut bank that was probably very close to the spot where Brown had first located fossil spruce cones and mastodon tusks. At first I saw nothing more than a pile of leaves, twigs, and other plant debris preserved in the bank where many years of stream action had eroded part of the fossil-bearing clay lens. Little Bayou Sara had thus stripped away some of the original exposure but had also opened up new portions of it. The blue-gray clay was not continuous for more than fifteen meters along the stream. Rather, lens-shaped pockets of clay represented what must have been oval pools scoured out of the former stream bed. Little Bayou Sara must have swept up and concentrated these materials in an ancient storm. They were then locked deep within a deposit of sand and silt, with only the bamboo-like stalks of American cane and a few horizontal, flattened logs emerging to show that the deposit continued back far into the cliff (Figure 13).

In the late Pleistocene, between ten and twenty thousand years ago, during dry seasons those pools must have been isolated from much of the stream flow. Stagnant water would have allowed silts and clays to settle out from suspension in the bottom of each pool. Leaves, nuts, twigs, conifer cones, and pollen grains floated into and then settled to the bottom of the pools, which eventually filled up with organic-rich mud. The muddy layer of blue-gray clay formed a protective coating that sealed off and helped to preserve fragile leaves and other macroscopic plant parts from decomposing. Animals coming to drink at the water's edge occasionally may have become mired in the sticky clay, caught in a flash flood, or killed by predators and perished, unable to extricate themselves. Their remains would have been preserved along with those of the plants because they would have been buried quickly in the clay. Each clay lens as we observed it consisted of mud that originally accumulated within a scour pool over as little as a year and at most a few hundred years — a very short interval in geologic time.

Figure 13. Paul Delcourt at stream-cut bank exposed in 1973 near Percy Bluff on Little Bayou Sara, West Feliciana Parish, Louisiana. (Photograph by Hazel R. Delcourt)

Little Bayou Sara exposed these fossils to the world by eroding a stream bank that continually caved in even as it was being undercut by flowing water.

81

As Paul and I removed shovels full of blue-gray clay from the bank, we uncovered cones, twigs, and needles of spruce. Our excitement increased because this was immediate evidence of the antiquity and importance of the Percy Bluff deposit — and because it corroborated Brown's original findings. We documented the occasion with a photograph of Paul posing in front of Percy Bluff, taken from a vantage point on a brush pile on the other side of an active scour pool filled with quicksand (Figure 13). Then we returned to the Land Cruiser, our backpacks filled with heavy, wet, blue-gray clay. We were wet and muddy and very tired from the day's stream-walking and excavations, but we were pleased and excited by what we had just found, and we were elated in anticipation of the discoveries that yet lay ahead.

Paul was keenly interested in documenting the plant macrofossils — the cones, needles, leaves, seeds, and wood — from Percy Bluff. He had been careful to take separate samples in a vertical sequence through the geologic deposits. We had labeled the samples in a series so that later we could arrange them from bottom to top, oldest to youngest, in the original sequence in which they were laid down by Little Bayou Sara. To separate the macroscopic plant fossils from the blue-gray clay, Paul placed a sample of clay in a beaker with a little hydrogen peroxide — enough to gently bubble apart the clay but not enough to damage the fossils. Then, carefully, he ran warm water over the clay, swirling it in the beaker, and pouring off the liquid into a standard sediment sieve. The sieve was a squat brass cylinder with a 250-micrometer wire mesh bottom designed to catch the plant fossils.

It was important to wash away the clay but not to break any of the small, delicate plant fragments. Paul saved as many bits of plants as possible so that he could identify them and also count their numbers. Carefully he picked out each one using a dissecting microscope, tweezers, and the tip of a fine camel-hair paint brush. Gingerly he lined up the plant remains in small plastic boxes and labeled them. He preserved them in a liquid mixture of glycerine, water, and formaldehyde.

When all the fossils were sieved, picked out, arranged in rows, and counted, the results, which included Paul's rediscovery of spruce remains, were very similar to the lists in *Bulletin 12*. Paul not only had found spruce cones, as had been reported earlier by Clair Brown, but also needle-like spruce leaves, twigs, and even delicate male cones, or microsporophylls, that still contained pollen grains. The assemblage of both durable and delicate white spruce plant parts was remarkably complete — an outcome that was not very likely if the spruce remains had floated hundreds of kilometers down the Mississippi River before coming to rest in Little Bayou Sara, as some had suggested previously. More likely, white spruce trees had grown on the banks of Little Bayou Sara when mastodons roamed the Tunica Hills.

The spruce fossils, however, weren't typical of any living species. They were fully twice the length of the cones of white spruce trees living north of Lake Superior. The dimensions of the spruce cones pictured in *Bulletin 12* (Figure 14) were similar to those Paul had found. The cell structure of the fossil spruce wood was also peculiar. The growth rings were abnormally narrow, which had led Brown to suggest that these trees grew slowly under stressful conditions. If the species of spruce was the same white spruce familiar to travelers across eastern Canada today, then the white spruce trees that reached the Tunica Hills

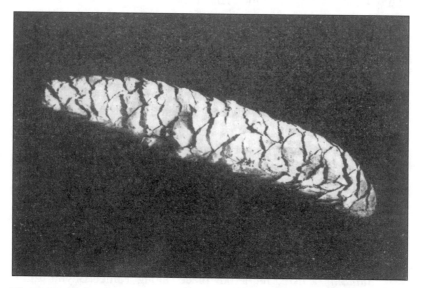

Figure 14. Spruce (*Picea*) cone collected by Clair Brown from Pleistocene deposits at the Percy Bluff fossil locality on Little Bayou Sara, West Feliciana Parish, Louisiana. This cone is 6.8 cm long. (Photograph reproduced from Fisk *et al.*, 1938)

during the Pleistocene must have been growing at the extreme limits of their physiological tolerance.

The spruce cones were a major puzzle — they just didn't fit in with the rest of the plant-fossil assemblage, which was mostly of temperate deciduous trees. Nor did they match descriptions of cones collected from living white spruce trees. They were much larger and had subtle differences in shape and in the dimensions of the cone scales. Further, the Tunica Hills spruce cones didn't match any other species of spruce living in eastern North America. Paul thought over the possible interpretations. One plausible explanation was offered in a study by Canadian scientists Hills and Ogilvie, who had discovered very similar fossil spruce cones in Pliocene-age deposits on Banks Island in arctic Canada. Their fossil species, *Picea banksii*, was thought to have become extinct more than two million years ago. Could Brown and we have discovered the last living enclave of this ancient tree, lost during much of the Pleistocene in the southeastern United States and found again in the Tunica Hills over twelve thousand years ago, only to be extinguished at the end of the Pleistocene a few thousand years later?

As an alternative to the hypothesis that the Tunica Hills spruce was an extinct species, we made our own collections of white spruce cones from living white spruce (*Picea glauca*) trees along the north shore of Lake Superior. We found that along Lake Superior, where summer temperatures remained cool and fog was persistent, the spruce trees produced cones that were bigger than those from trees growing inland. We thought that the Tunica Hills spruce might well have been an ecotype, a race, or a subspecies of *Picea glauca* that was eliminated from the southeastern United States at the transition from Pleistocene to Holocene climatic conditions. Without more data, Paul preferred to think that the Tunica Hills spruce was a variety of white spruce that had become extinct at the end of the Pleistocene, rather than the less likely possibility that it had been a relict of the Tertiary Period, stranded for millions of years in an isolated location in the Tunica Hills.

Bulletin 12 listed numerous places along Tunica Bayou, Little Bayou Sara, Big Bayou Sara, and Thompson Creek where mammal fossils had been collected. But rarely were the fossil bones found as intact skeletons. Instead, the fossil bones of Pleistocene mammals were generally found strewn about in the stream beds, washed out of the stream banks and concentrated by water action according to their size and density in deposits of cobbles or in sand bars in the middle of the stream bed. Most of the fossil horse and mastodon bones could not be dated accurately because they had been removed by the stream from deposits in which they were initially buried. The plant fossils, however, were still to be found within stream terrace clays.

But were the blue-gray clay layers along all the bayous deposited at the same time? Or were there many different blue-gray clay layers, each contained within the ancient floodplain deposits of a stream terrace dating from a different period? We had two ways of determining the age of the blue-gray clay. The first was by radiocarbon dating, which involved collecting fossil-laden sediment, washing and sorting pieces of plant matter, and sending them to a radiocarbon dating lab for analysis at a cost of several hundred dollars per sample. LSU had such a lab, and Paul learned how to date fossil wood himself. His radiocarbon dates on wood taken from the blue-gray clay layer at Percy Bluff, and containing both spruce remains and fossils of deciduous forest trees, confirmed that the mixture of boreal and temperate species had coexisted during the late Pleistocene, about twelve thousand five hundred radiocarbon years before present.

The second way of dating the fossils, a technique particularly useful for fossils too old for the radiocarbon method, was by drawing geologic maps and cross-sections of fossil-rich terrace exposures. If we could map the stream terraces, we could determine their relative ages based on their elevation and position along the streams. These maps would let us see how the stream deposits of Little Bayou Sara compared, or correlated, with similar deposits that had been mapped along other rivers across the whole Gulf Coast region. We decided that next we needed to understand the geology of the stream terraces as a context for understanding the significance of the plant fossils.

At first, mapping the terraces seemed straightforward, a simple extension of previous work by Fisk. During the 1930s, Fisk had determined elevations using a survey aneroid, which operated on the principle that the barometric pressure of the atmosphere changes with differences in altitude. He calibrated the instrument each day by determining the barometric pressure difference in the elevator in the LSU geosciences building. After calibrating the aneroid, Fisk and his graduate students would drive north from Baton Rouge, taking an aneroid reading every thirty seconds to calculate changes in elevation of the land surface. Some of the elevation determinations could be checked with official United States Geological Survey benchmarks. But the aneroid instrument was sensitive of changes in barometric pressure — it was influenced not only by changes in elevation of the land surface but also by daily convection thunderstorms. Its readings drifted during the day, and by afternoon the elevation numbers were different from those that would have been recorded earlier in the morning. By then the LSU geosciences elevator was too far away to be used to recalibrate the equipment.

We discovered this problem by comparing Fisk's maps with more recent topographic maps produced by the United States Geological Survey using modern cartographic methods, refined field measurements, and aerial photography. Paul found that when he projected the elevations of Fisk's terraces on the modern map, two of Fisk's four proposed terrace surfaces would have to be poised in mid-air, high above the true elevations of the highest land surfaces in the Tunica Hills.

Fisk thought that the land surface of each terrace was located at the highest position sea level had reached in the nearby Gulf of Mexico during a particular time of warm interglacial climate. Periods when the sea level was high were followed by periods when it fell as moisture evaporated from the oceans and was stored on land as ice during times of cold glacial-age climate. When the sea level was low because water was locked away in frozen glaciers, the Mississippi River and all its coastal tributary streams cut down through their floodplains to reach the new base line, abandoning terraces composed of former stream deposits. These terraces were then perched along the flanks of each river valley. Each terrace thus represented one cycle of sea level change, and each was produced during one complete cycle of interglacial and glacial climate.

Fisk thought of southern Louisiana as pivoted on the balancing hingeline of a gigantic teeter-totter. Mud dumped by the deltas of the Mississippi River caused the floor of the northern Gulf of Mexico to subside in a collapsing depositional basin that dropped down as much as several kilometers during the Pleistocene. As the Gulf of Mexico coastline sank downward, inland in the Tunica Hills the land rose as a Southern Mississippi Uplift. According to this explanation, the oldest terraces would have been uplifted for the longest time and therefore would have the highest elevations. Progressively younger terraces would then be found at successively lower elevations and closer to the Gulf Coast.

Fisk assumed that there were four sets of coastal and stream terraces across southern Louisiana, each corresponding to one of the four great intervals of Pleistocene glaciation then thought to have occurred in the northern hemisphere. This conclusion, shaped by the state of knowledge before the advent of radiocarbon dating as an absolute dating technique, and based in part on geologic evidence from midwestern glacial deposits, was later revised. In the midwestern United States, only the most recent ice sheet advances can be mapped in detail — erosion has removed much of the evidence of previous ones. This means that the geologic record on the continent leaves critical parts of the story blank. Only in ocean basins is the complete record of Pleistocene history intact, contained in marine mud that has been deposited continuously on the ocean floor.

In the 1960s and 1970s, Quaternary geologists began to examine cores of ocean sediment. They discovered that, rather than only four cycles of advance and retreat of land-based glaciers during the Pleistocene, more than twenty glacial-interglacial cycles have occurred during the past two million years.

Moreover, through the millennia streams cut down through the coastal plain toward their local base levels. For streams and bayous of West Feliciana Parish, this base level was the elevation of the Mississippi River. Over time, intervals of down-cutting of the upland terrain have alternated with periods during which streams filled in their valleys. Over long cycles of changing climate the streams fluctuate up and down in response to changes in rainfall, sea level, or local base level. Valley filling followed by more down-cutting eventually produces a series of terraces that represent former floodplains that are now abandoned. These terraces are consequently higher than the present stream floodplains but adjacent to them. Sediments of each terrace accumulate during an aggradation phase; gravel drops out of the stream first, followed by sand and then silt, which are both less dense and smaller in particle size than the gravel.

The events along Little Bayou Sara have been linked to the history of the Mississippi River and to changes in water and sediment flow related to the growth or retreat of great continental ice sheets. During the Pleistocene, when glacial meltwater flowed south from the margin of the continental Laurentide Ice Sheet, the Mississippi River was deeply incised within its valley and was a braided river, with many intertwined channels spreading out and coalescing across its wide valley. Westerly winds lifted silt off the exposed river bars in the braided-stream channels and swept the silt eastward, where it was draped over the bluffs along the eastern edge of the Mississippi River valley. Stream incision and infilling along Little Bayou Sara also have been influenced by changes in the level of the Gulf of Mexico, which has risen and fallen as much as 120 meters over the course of each glacial-interglacial cycle.

When we realized that Fisk's classic mapping of terraces in the Tunica Hills needed complete reinterpretation, Paul and I took to the field with our survey staff and inclinometer. Our aim was to draw cross sections across the major streams at key points tied into United States Geological Survey benchmarks and right-of-ways surveyed by the Louisiana Department of Transportation. Paul could interpret the local geology from these cross sections to produce a more realistic geologic map than had been available before. In the process, we scaled near-vertical cliffs and scratched our way through living barbed wire of cat brier (*Smilax*) vines and bramble (*Rubus*) thickets.

Our terrace mapping culminated at a site we called the amphitheater cut, which turned out to be the key location at which the fragments of geologic evidence all fit together. The amphitheater cut is a big scalloped cliff carved out by flowing water. This huge natural meander of Big Bayou Sara is now abandoned by the stream but exposes a fifteen-to twenty-meter vertical wall with a complex series of stream deposits, all capped by a massive layer of loess. At the amphitheater cut, Paul found a stack of ancient soils with layers of sand and silt in between.

These fossil soils, called "paleosols," gave him a way of determining how many episodes of climate change and stream terrace building were preserved in the area, and this geologic section became the template for developing the relative chronology of geologic events for the late Pleistocene in the Tunica Hills.

Based on the geologic evidence from the amphitheater cut as well as from many other locations where we drew cross-sections and underlying stratigraphy across the floodplains of the principal bayous and streams of West Feliciana Parish, Paul revised Fisk's original geologic map (Figure 15). The lowermost of Fisk's terraces, the Port Hickey or Prairie, was split into two. The lower of these dated from about twenty-three thousand years ago to the present. The upper terrace was deposited during an interglacial interval some one hundred thousand years ago. The summits of the Tunica Hills, Fisk's Montgomery, Williana, and Bentley terraces, were reassigned to the coastal-plain sands of the Citronelle Formation, which formed several million years ago. Thus, we made another step toward understanding the history of the geology of the Tunica Hills.

From the revised terrace maps and the few radiocarbon dates we had on fossil wood, we knew that most of the blue-gray clay deposits that contained plant fossils in the Tunica Hills dated from the late Pleistocene through the early Holocene. In sediments dating from twelve thousand five hundred radiocarbon years ago, Paul had found both abundant remains of boreal spruce and fossils of cool-temperate deciduous forest species. Plant fossils representing deciduous forest elements included beechnuts, winged seeds or samaras of tulip tree, and fruits of hornbeam. Acorns of warm-temperate oaks and grassy culms of American cane were preserved in the younger Holocene deposits. We concluded that during the late Pleistocene, a mixture of white spruce and deciduous forest co-existed in the Tunica Hills. We reasoned that, since white spruce had migrated as far south as West Feliciana Parish, it probably had existed all along the Blufflands during the late Pleistocene.

We couldn't answer Deevey's question of whether deciduous forest had been pushed far to the south into peninsular Florida or the Valley of Mexico. We could, however, suggest, on the basis of the plant-

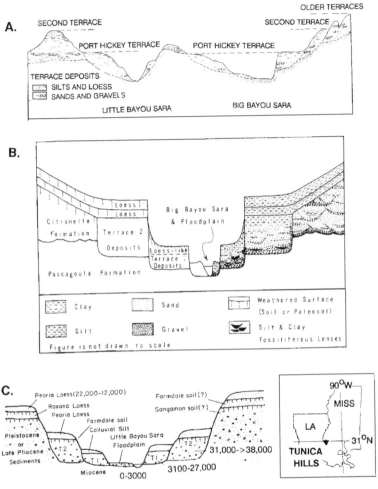

Figure 15. Three interpretations of the stream terrace formation in the Tunica Hills of southeastern Louisiana. **A.** Fisk (1938) **B.** Delcourt (1974; Delcourt and Delcourt, 1977) **C.** Jackson and Givens (1994). (Diagram from Delcourt and Delcourt, 1996)

fossil evidence from the Tunica Hills, that deciduous forest communities had persisted in at least one area of the Gulf Coastal Plain through the Pleistocene. We could also state unequivocally that climate change associated with glacial advances in the north had affected the distributions of at least some species of plants south of the glacial margin during the Pleistocene, in contrast to Braun's assertion that no changes in biological communities had occurred.

90

Reality, we thought, lay in between the two extreme viewpoints. But if there were neither wholesale southward migration of species nor stasis, what kind of climate changes could account for the unusual intermingling of northern and southern species in the late Pleistocene in the Tunica Hills? Following from our experience in living on the flood-plain of the Mississippi River and from collecting northern plants along the Blufflands in deep, moist ravine habitats carved into the loess, we formulated a hypothesis about long-term climate change in the Tunica Hills.

From our studies in Michigan, we had learned that a number of plants there lived south of their general limits of distribution along the shores of the Great Lakes. Tundra plants such as crowberry (*Empetrum nigrum*) lived on rocky terraces above Lake Superior; jack pine and its northern associates grew south nearly to Chicago along Lake Michigan, but not inland. These southern range extensions are possible because of the lake-effect climate — ameliorated climate within just a few kilometers of large bodies of relatively cold water. Could a similar cold-water influence have created suitable microclimate in the Blufflands during the Pleistocene?

We had observed that in late winter and spring, the water in the Mississippi River was relatively cold even as it reached Baton Rouge. At that time of year, the Mississippi was largely fed by melting snow from its tributaries in the upper Midwest and the Northeast. Air temperatures along the Gulf Coast, however, are often balmy even in January and February, because of the influence of the maritime tropical air mass emanating from the Gulf of Mexico. The clash of the cold river water and the warm air often produced persistent fog through the process of advection of moisture from the river surface into the air.

What if, instead of being fed by cold water for only part of a year, the Mississippi River was fed a constant stream of cold water throughout the year? During the late Pleistocene, when the Mississippi was a braided river, its many hundreds of active channels were choked by silt carried by meltwater issuing forth from beneath vast continental ice sheets. During the late Pleistocene, and especially during the late-glacial interval from about ten to eighteen thousand radiocarbon years ago, both the Cordilleran Ice Sheet over the northern Rocky Mountains and the Laurentide Ice Sheet centered over Hudson Bay were thinning from their maximum domes. The eastern ice dome was four kilometers thick but was beginning to melt and to retreat northward from its maximum southern extent near the location of modern St. Louis, Missouri. Much of the meltwater was funneled down the Mississippi River Valley. Wind-blown layers of silty loess formed during much of the Pleistocene, but the thickest blanket accumulated during the time at which the last great ice sheets reached their maximum extent, during the full-glacial interval eighteen to twenty-five thousand radiocarbon years ago.

We surmised that during the last full-glacial and late-glacial intervals, when global climates were coldest and continental ice sheets were at their maximum extent, cold meltwater constantly funneled down the Mississippi River, past the Blufflands and the Tunica Hills. The glacial meltwater would have come into contact with relatively warm air over the Gulf Coastal Plain, and persistent advection fog could have been characteristic much of the year. The advection fog would have produced a local ameliorating effect on the climate of the Blufflands region, allowing for southward range extensions of some boreal species such as white spruce. The Pleistocene Blufflands climate would also have been favorable for cool-temperate species, such as the assemblage of northern plants that were thought to have been disjunct in the Tunica Hills before we collected them throughout the Blufflands.

Thus, the Blufflands may have served as a migration pathway for species adjusting their ranges to climate change during the Pleistocene and the Holocene. Persistent advection fog would have allowed for coexistence of northern and southern species in biological communities unlike any we find today. Our Blufflands migration hypothesis allowed for a middle ground between the two extreme hypotheses of Braun and Deevey.

But our Blufflands hypothesis was still based upon scanty evidence from a handful of fossil collections in the Tunica Hills and the modern plant collections from the Blufflands. Having begun to solve mysteries in the Tunica Hills, we became acutely aware of the need to expand our research and to develop a systematic and comprehensive search for the elusive Pleistocene deciduous forest. First, could we confirm our Blufflands hypothesis by finding Pleistocene spruce and deciduous forest elsewhere in the Blufflands region? Second, was the advection fog idea valid, or was there a regional change in climate that caused widespread adjustments in species distributions both east and west of the Mississippi River? Did spruce grow elsewhere on the Gulf Coastal Plain, or was it confined to the Mississippi River Valley? Where were deciduous forest communities during the Pleistocene?

Our fossil finds in the Tunica Hills remained as controversial as those of Clair Brown for some time. Then, in the 1980s and 1990s, additional work was done by professors Stephen Jackson of the University of Wyoming and Charles Givens of Louisiana State University. Jackson and Givens, fortified by funding from the National Science Foundation, revisited the Tunica Hills and excavated fossil spruce remains from a number of localities along the major streams and bayous, and their tributaries. Their results added substantially to the base of information, not only confirming the original finds but also extending the record of spruce back to full-glacial times with many radiocarbon dates of twenty thousand years before present or greater. From the full-glacial deposits, Jackson and Givens also found macrofossil remains of deciduous forest species, including walnuts and acorns (figures 15, 16).

Based largely on measurements of spruce cones and spruce seeds from the Tunica Hills, Jackson concluded that the fossil spruce was not the modern white spruce (*Picea glauca*) that today is widespread in Canada. Further, the Tunica Hills spruce was not only different from any modern species, but it was also distinguishable from *Picea banksii* and therefore was not a relict from the Tertiary Period. Jackson named the now-extinct Tunica Hills tree species "Critchfield spruce" (*Picea critchfieldii*) in honor of the eminent Canadian spruce taxonomist William Critchfield. Another puzzle remains, however, as even in the 21st century Quaternary paleoecologists continue to speculate on the climatic and ecological significance of the extinct Tunica Hills spruce.

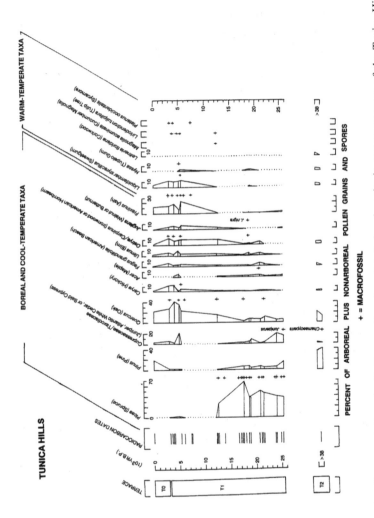

Figure 16. Composite diagram of plant fossils from radiocarbon-dated clay deposits in stream terraces of the Tunica Hills, Louisiana-Mississippi. This diagram is based on the data of Delcourt and Delcourt (1977) and of Jackson and Givens (1994). (Diagram from Delcourt and Delcourt, 1996)

94

Expanding Horizons : The Stories that Fossil Pollen Tell

The pond lay still, as if anticipating. Its darkness mirrored the encircling spruce trees. Their reflected images grew dim in the gathering dusk. Water lilies (*Nuphar* and *Nymphaea*) covered the firm, cool lake bottom with a profusion of tangled stems and roots, their leaves gracefully floating at the pond's surface. Only an occasional ripple broke the stillness. A mastodon, browsing nearby, approached the water's edge to drink, inhaling tea-colored water and releasing the odors of decaying leaves and swamp forest floor. Southern bog lemmings (*Synaptomys cooperi*) scurried almost noiselessly through the stalks of ripening wetland grasses, searching for seeds to eat. Above, a snowy owl (*Nyctea scandiaca*) began its nocturnal vigil. A bald eagle (*Haliacetus leucocephalus*) paused in the gnarled limb of a large spruce tree overlooking the pond, then spread its wings and wafted away. Blackburnian warblers (*Dendroica fusca*) flickered, black and orange in the treetops. In the hills beyond, the evening breeze ruffled aspen and birch leaves. The plateaus beyond the pond seemed infinitely evergreen, clothed with a forest of jack pine.

A bolt of lightning and a crack of thunder split the silence. The late-summer storm blew up quickly. The first raindrops turned to torrents and the squall line passed over the plateau and descended into the valley. Rivulets filled, then overflowed into streams. Silt and leaf litter journeyed with the swirling water down the hill slopes to the awaiting pond. Torn from branches in the gale, spruce and jack pine needles drifted down into the pond. Gradually becoming soaked, they slowly sank to the bottom. Soft mud buried them. They joined the myriad plant and animal bits entombed there — algae, water fleas, water lily seeds, pollen grains, and fern spores. Nearly every living thing in or near the pond contributed materially to the accumulating layers of mud.

Steep slopes surrounded the pond, whose shape reflected its origin within limestone rock leached deeply by the rainfall of tens of thousands of years. Many sinkhole ponds dotted the valley floor. Marshlands formed behind dams created by giant beavers (*Castoroides ohioensis*), interconnecting the ponds. Fragrant white cedar (*Thuja occidentalis*) and spruce trees dominated in swamps on the slightly higher ground that marked the edge of the open, grassy marsh. Between the cedar swamps and the distant jack pine-covered sandstone plateau, steep slopes harbored a deciduous forest of birch, maple, and oak.

This was the landscape of a valley in Tennessee twenty thousand years ago, as I imagined it, based on vegetation reconstructed from my counts of fossil pollen grains (Figure 17). Identifying and counting fossil pollen grains is a process that has always transported me into another dimension. Once I set up my sample slide and find a comfortable position at my desk, I fade off into a trance-like state during which only the scenario unfolding within the microscope system matters. Peering into my microscope, I feel as if I am deep within the full-glacial boreal forest of Tennessee. As I walk among the trees of the open forest, I am keenly aware of the balsam and piney scent of the evergreens and the crunch of dry jack pine needles beneath my boots. I hear a flutter of wings — could that be a Kirtland's warbler (*Dendroica kirtlandii*) flitting by? Here the forest floor is only sparsely vegetated, with a little bracken fern (*Pteridium aquilinum*) beneath the canopy of jack pine. Oh, but what is that? A meadow rue (*Thalictrum dioicum*). Over there, beyond the far jack pine tree, I see a thicket of hazelnut (*Corylus americana*). At the forest edge, I can just make out a cluster of blueberry shrubs. Sedges (Cyperaceae) flourish at the margins of ponds and in wet meadows along streams. I stop still in my tracks, fascinated by a plant so new to me that I know I must examine it carefully and look it up in my identification manual. I need to take time out, to sit down next to that plant and find out what it is. I feel slightly disappointed, because it was such a nice day and I had been making such good time — I wish I could have completed the scene.

As I emerge from the mental hologram, I reach for my manual for identifying pollen grains. The unknown microscopic object turns slowly, bathed in silicone oil between the cover slip and the microscope slide. It reveals intricate patterns of sculpturing on its surface and six distinct,

Figure 17. Lake in northeastern Minnesota that serves as a modern analog for full-glacial environments at Anderson Pond, Tennessee. (Photograph by Paul A. Delcourt)

delicate furrows with pores punched out of their centers. I identify it as a burnet (*Sanguisorba canadensis*), a relatively rare plant in the fossil pollen assemblage I had been working on. Burnet is a small boreal herb, a member of the rose family, with lacy leaves and a tuft of pink flowers. Returning to my trance, I see that I am close to a clearing in the forest. A small bunch of burnet appears in the hologram precisely at the point where I had stopped, at the edge of the pond.

The starting point for recording each pollen spectrum is a blank tally sheet. For each pollen grain that I see in a microscope slide and tally on my score sheet, I imagine a plant growing on an ancient landscape. Gradually the paleo-landscape becomes populated with plants growing in wetlands, on hill slopes and in ravines, and on uplands. Each microscope slide represents a primeval landscape captured at one moment in time. In my mind's eye a picture unfolds and stands out in bold relief like a museum diorama, a still life in which each individual plant has its own special place. By adding a second sample, representing a different slice of time following the first, I can compare them, noting changes in the composition of plants and changes in their abundance. Some species disappear and are replaced by others through this panorama of time. With a large set of samples, I can begin to visualize the entire course of events, as if I had stepped into a hologram and become part of the virtual scene. Secreted behind their microscopes, for decades Quaternary paleoecologists have thus tallied fossil pollen grains and conjured up landscapes of the past.

Pollen grains have been called the perfect fossils. They are a nearly ideal type of fossil assemblage for interpreting changes in vegetation and hence in climate. Most pollen grains of temperate trees are produced abundantly and transported easily by air currents. When mixed into lake water and settled to the bottom of a lake or bog, they are incorporated quickly into the muddy bottom sediments and fossilized easily. Because pollen grains tend to be well mixed and ubiquitous through the environment where they are produced, palynologists can use counts of fossil pollen grains to develop a statistical fingerprint for interpreting the type of vegetation that produced the pollen spectrum.

Correct identification of pollen grains and spores is the cornerstone of Quaternary palynology. When Paul and I decided to expand our search for the Pleistocene deciduous forest beyond the Tunica Hills, we knew that we needed a larger tool kit. We needed to be able to analyze fossil pollen assemblages from samples we would take from lake sediment from a variety of sites in the southeastern United States. We also knew that the credibility of our research results depended in great part on the accuracy of our identifications of fossil pollen grains, as well as our ability to identify plant macrofossils.

While at LSU, Paul and I visited a number of botany departments across the southeastern United States to collect pollen as well as fruits and seeds from herbarium specimens. When we moved to the University of Minnesota in 1974 to study fossil pollen analysis with Herb Wright and Ed Cushing, we divided these collections of southern plants and traded portions of them for northern species from the collections of the University of Minnesota's Limnological Research Center. These reference collections have been a mainstay of our work and that of our students over many years at the University of Tennessee.

Pollen grains are sometimes called flower dust. They are tiny, balloon-like capsules that protect male sperm cells from drying out on their precarious journey from flower to flower. The tough cell wall of the pollen grain has evolved through millions of years to become relatively thick and complex in structure. The pollen wall, or exine, is usually built up from three layers, very much like a Greek temple. It has an inner layer that forms a foundation, an outer layer or roof-like tectum, and vertical supports called collumellae that support the roof-like outer cell wall. Once the male pollen grain lands on a receptive female surface of a flower of the same species, the grain germinates. It sends down a tube through which the sperm cell migrates to fertilize an egg in the ovary of the flower. Each pollen grain has one or more weakened, thin spots in its cell wall through which the pollination tube can grow. If these weak spots are round, they are called pores; if they are long gashes, they are called furrows.

Pollen grains are smaller than grains of beach sand, and they come in a fascinating variety of shapes and with many different patterns sculpted on their outer surfaces (Figure 18). Oak pollen grains look like miniature deflated basketballs, with three broad furrows slashed in their sides and an irregular bumpy texture covering the outside. Maple pollen has three slit-like furrows and looks like a skein of yarn, with delicate strands crisscrossing its finely spun surface. Eastern hemlock pollen looks like a piece of shag carpet wrapped around a tennis ball.

99

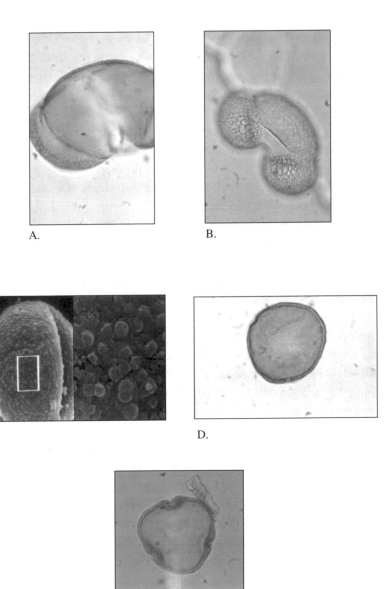

A.

B.

C.

D.

E.

Figure 18. Representative kinds of pollen grains typical of boreal and temperate forest. **A.** *Picea glauca*. **B.** *Pinus banksiana*. **C.** *Quercus*. **D.** *Carya*. **E.** *Tilia*. (Photographs A, B, D, and E by Paul A. Delcourt; photograph C by Allen Solomon)

Hickory pollen is more like a billiard ball, completely smooth on the surface, and punctured with three circular pores nearly evenly spaced around its perimeter. Ragweed pollen looks like a medieval torture device. It is a ball covered with long, sharp spines. Pine pollen looks like a Mickey Mouse hat, with a central cap and two large, inflated bladders forming ears on either side. These air-filled bladders keep the otherwise dense pine pollen grain buoyant so that it remains airborne during pollination. In pine pollen grains, the boundary between Mickey's ears and his head is a sharp line. The bladders attach to the cap like a French shopping bag of net mesh drawn in with a string. In spruce grains, which are also much larger than pine pollen, the boundary area between the side bladders and the central cap is much more gradual. Balsam fir (*Abies balsamea*) pollen is the largest and most robust of all the conifer grains, with a very thick frill along the top of the cap where it meets the bladders.

The pollen grain features of many groups of plants are quite similar to each other. Finer and more subtle distinctions have to be made in order to differentiate and identify them to genus or species. Sculpture patterns on the outer surface range from low bumps to high spines, and from a netlike pattern to swirling strands. Pores and furrows differ both in number and in their internal structure. For example, birch, hazelnut (*Corylus*), and wax myrtle (*Myrica*) pollen are all small, smooth grains with three pores. They can be distinguished from each other by their pore structure. In birch grains the cell wall is puckered up around each pore so that from the surface the pore looks like a short spout. In cross section, when viewed through the pore from one side, the pore is aspidate — that is, the flaps of the cell wall curve upward and flare out to form the edge of the bulging pore. They look like the heads of two cobra snakes about to strike each other. The pore structure of the other three-pored pollen grains is more subdued. In hazelnut, the inner cell wall stops short of the base of the pore and the outer wall extends over it. In wax myrtle, the pattern is similar to hazelnut except that a few flecks or rough bumps of inner cell wall extend beyond the place where the cell wall stops.

To make a slide that shows the characteristics of pollen grains, one must first collect flowers from living plants at the height of their bloom. After collecting pollen from the plants, the pollen must be processed to render it into a form that can be compared directly with fossil pollen grains. Chemical processing is a painstaking procedure that first involves washing the pollen grains off of the flower and through a screen with running water. Then the grains are heated for thirty seconds in a small test tube containing nine parts acetic anhydride to one part sulfuric acid. This seemingly caustic treatment, called acetolysis, consumes the cell contents of the grains but leaves their outer wall in perfect condition, transparent and light brown, so that not only the surface sculpture patterns are visible, but also the inner structure of the cell wall can be seen. The acetolyzed pollen becomes very similar in appearance to fossil pollen grains retrieved from bog or lake mud. The residue containing pollen grains is suspended in thick, clear silicone oil which has the viscous consistency of honey. Then a drop of this mixture is mounted on a glass slide, trapped underneath a sealed glass cover slip, and labeled. These microscope slides have a shelf life of only a few years to a decade before the mounting medium cracks, leaks, or chemically reacts with the pollen grains in the residue. Because of the short life of the microscope slides, the balance is archived in vials for long periods of time so that the slide collection can be replenished as necessary.

Palynologists have compiled several reference books of common pollen types to aid in making routine identifications of fossil pollen grains. The pollen types, in the form of photographs or drawings, are arranged according to their identifying characteristics. Similar types are distinguished by their one major difference, and so on down the line, with the use of a dichotomous, or branching, identification key in which each question about the appearance of the grain has two possible answers. Does it have pores or furrows? If there are pores, are there one, two, three, or more of them per grain? If there are three, are they arranged symmetrically, evenly spaced around the perimeter, or are they offset, with two on the perimeter and one facing you head-on? Hundreds of choices involve the kind and arrangement of pores, the thickness and construction of the cell wall, the type of sculptural pattern on the surface of the grain, the shape of the grain viewed from several different orientations, and so on.

Once the characteristics of the common pollen types are committed to memory, finesse comes with being able to identify the occasional "unknown" pollen grain. The challenge is in developing the necessary and sufficient repertoire of known pollen types for deducing the identity of those that occur only rarely in fossil sediment.

Paleoecological articles typically include pollen diagrams that are printed as oversize pages, most often bound into journals as foldouts, but sometimes included as appendices in a pocket glued to the inside cover or printed as page-sized summary diagrams. The full diagrams are as expensive as maps to produce and to print. They are large because of the great amount of data represented on them. For a given fossil pollen site, fifty to one hundred pollen samples may be counted, with five hundred to two thousand individual pollen grains tallied per sample. As many as 150 types of pollen grains and fern spores may be identified throughout the course of a fossil record that spans enough time to reflect major changes in climate and environment. On the pollen diagram that summarizes the results of pollen analysis, each pollen type is represented by a vertical curve depicting changes in its abundance throughout the core sequence. Each curve is labeled with the scientific name of the plant that produced the pollen type. Finally, a scale is plotted on the diagram that shows depth, age based on radiocarbon dates, and changes in the texture and color of the sediment as well as the types and amount of organic matter it contains.

All this information helps make sense of the fossil record. How old is the lake mud? How rapidly or gradually have changes occurred at the site through time? What were the main constituents of the former vegetation? When was one type of vegetation, for example, tundra, replaced by another vegetation type, such as forest? What was the sequence of entry of different kinds of forest trees onto the landscape surrounding the bog or lake from which the fossil pollen sequence was analyzed? Is there evidence of human influences near the site — is there, for example, pollen of cultivated plants? None of these interpretations are obvious from thumbing through a three-ring binder stuffed full of score sheets, with each tally sheet representing an individual sample. All of these questions and more, however, can be answered at a glance from the complete pollen diagram. It is a challenge to summarize concisely a vast amount of information in a form that makes it easy to visualize landscape changes through time. The pollen diagram solves this problem, and it has been the standard form of presenting fossil pollen data since 1916, when Von Post published the first paper on pollen analysis as a means of interpreting forest history.

Between collecting cores of lake sediment and publishing the final diagrams lies a vast expanse of lab work and data analysis. Preparing pollen residues from raw sediment involves a long series of chemical treatments, which is much more complex than making modern reference slides from herbarium specimens. First, potassium hydroxide, a strong base, is used to remove the humic acids produced by the decaying plant matter. Then, strong acids such as hydrochloric acid and hydrofluoric acid are used to remove calcium carbonate (lime) and silicates (quartz), respectively. Acetolysis is used to render the pollen grains transparent. The pollen residue is then dehydrated in alcohol and transferred to storage vials containing silicone oil. After each step, the remaining material is centrifuged down and the waste products of reacted chemicals are decanted off. Each sample of mud begins as only a thimble-full, and so very small quantities of each reagent are used. Even so, each mud sample may contain a million or more fossil pollen grains.

An experienced lab technician can prepare a dozen samples of lake mud in a day. For each study site, a week to ten days of lab work is required just to render the mud samples into usable form. Processing sediment samples with caustic chemicals requires good dexterity, patience, and care not to spill any toxic substances. The most dangerous of the chemicals is hydrofluoric acid (HF). Even in very small quantities, HF can be deadly, and a few parts per billion in the air can etch glass on the fume-hood window. If absorbed through the skin, HF will eat bone. It is volatile at normal room temperature, so the pollen extraction lab must be kept cold to keep the acid from turning into a harmful gas. When treated with due respect, including neutralizing any left-over acid, this and all other chemicals used routinely in the pollen lab make it possible to recover information about climate and vegetation history otherwise lost on the bottom of lakes and bogs.

Counting pollen requires making an objective and systematic scan

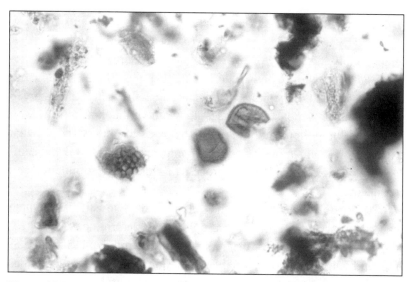

Figure 19. Pollen residue as seen through a microscope at 400X magnification. (Photograph by Hazel R. Delcourt)

of a representative slide of pollen residue (Figure 19). At a magnification of 400 to 1000 times actual size, a record is made of each fossil grain encountered on parallel but non-overlapping traverses of the slide. Traverses are made by moving the microscope stage front to back underneath the lens, then offsetting by a small interval and moving the stage front to back again. This process is repeated until half or a whole slide is canvassed, with care taken not to count an individual pollen grain or spore twice. On a routine slide, a count of five hundred pollen grains and spores can be accomplished in two to four hours. A systematic effort is critical to accurately describe the contents of the fossil sample and to have a statistically reproducible result — both the edges and the center of the slide must be included, because the larger pollen grains and spores tend to settle out toward the center of the slide and the smaller ones tend to roll toward the outer margins of the cover slip.

Pollen grains that can't be identified are placed in one of two categories: unknowns and indeterminables. Unknown pollen grains are puzzles that cannot be solved with the use of any standard reference book. They are morphological types that the pollen analyst has never seen before. They remain unknown until identified by finding an exact match with reference material. Indeterminable pollen grains are those that literally can't be determined, no matter what the extent of experience, because they are broken or crumpled or deteriorated beyond recognition as to specific type. Indeterminable pollen grains must be declared in the final tally, however. Even though they can't tell us much about plant life on past landscapes, they are an index as to how well preserved the pollen sample is. This helps us to understand to what extent information about the original composition of the pollen assemblage may have been lost in the process of sediment accumulation and fossilization.

Ultimately, the age range and sediment characteristics with which the Quaternary paleoecologist has to work are determined by the geographic location, geologic origin, and the area and depth of a lake or bog study site. The pollen analyst must adjust his or her techniques according to the constraints of the particular site. In Minnesota lakes, the rule of thumb is that mud accumulates on lake bottoms at a more or less constant rate of about one meter per thousand years. Depending on when the lake formed, the typical thickness of mud beneath a medium-sized kettle lake is about ten to twelve meters and represents the total accumulation of sediment over the past 10,000 to 14,000 years since the site was freed from melting glacial ice. Often pollen analysts use pollen time lines to gauge the rate of sediment accumulation in a site. These time lines are defined by changes in the abundance of particular types of pollen grains that are found in similar proportions in a number of sites. The changes in pollen assemblages reflect changes in vegetation that are presumed to have occurred at the same time over a broad region and hence can be used to date changes in a newly studied lake site. If the changes are synchronous and pinned down with an absolute radiocarbon date for the event at one site, then similar events and their time lines can be matched up from site to site and thereby help to date other events as well. This type of correlation method is a standard tool in paleontology, where it is often difficult to determine the true age of past events.

In the southeastern United States, there are no prescribed guidelines for lake development. Lakes vary widely in time of origin, and because there have been major changes in climate, both the type of sediment and the rate at which it is deposited in them vary through time. In Minnesota, the standard strategy for sampling sediment cores for pollen analysis is to take pollen samples every ten centimeters through the length of the sediment sequence. If the sediment accumulation rate has been constant through time, this results in samples evenly spaced in time, about every hundred years, as well as in depth. This sampling interval makes for a pleasing presentation on a pollen diagram and makes the interpretation of events through time much easier. If this assumption of uniform rates of sediment accumulation cannot be met, as is the case in most southeastern sites, then a different sampling strategy has to be employed. Preliminary radiocarbon dates are taken to establish the overall time range of the site. Changes in type of sediment are also dated, as they may reflect changes in climate that have affected the rate of sediment accumulation. Finally, preliminary pollen counts from widely spaced sediment samples can give an indication of the rate of change in vegetation through time and hence can guide the selection of depth interval for samples to be taken from the cores of lake mud to capture finely spaced events in vegetation history.

Reconstructing past landscapes from fossil pollen evidence is generally acknowledged to depend in large part upon the individual investigator's skill in pollen identification and experience in plant ecology. The more one understands the relationships of plants to their modern environments, the more likely one will be able to interpret correctly the paleoenvironments represented by fossil plant assemblages. The mental hologram approach to landscape reconstruction, however, only goes so far. Beyond determining which plants grew on a landscape in the past, generalizations about the abundance of those plants and their spatial relationships require interpreting percentages or absolute amounts of pollen found in sediment samples in terms of the relative or total abundance of plants living in the vegetation of the past.

How perfect are fossil pollen assemblages? How accurately do they reflect the vegetation that produced them? The quest for translating fossil pollen assemblages in terms of the quantitative attributes of the vegetation that produced them has been one of the major occupations of Quaternary paleoecologists over the past century. The quest for quantification has led to a host of related questions that have been the subject of intensive research investigations. Research into pollen representation, in turn, has led to the making of maps of past vegetation and climate that have become increasingly sophisticated over time.

A number of interrelated questions have arisen in the course of reconstructing landscapes from Quaternary pollen data. How is the pollen record influenced by conditions in which pollen is buried and preserved within lake mud? Does pollen deterioration within lake environments distort the fossil pollen record so that the percentages counted in a fossil sample are no longer the same as the proportions produced by the original plant community? Do some plants produce more pollen grains than others? Is the amount of pollen produced disproportionate to the abundance of a given species on the landscape surrounding a lake site? How closely do the percentages of pollen grains of different types correspond to the composition of the vegetation in the local area or the broader region around the lake? How is the pollen transported to the lake? How much of it is lost as it strikes tree trunks or shrubs rather than the lake surface?

Answers to some of these questions have become clear as European and North American paleoecologists have found ways to tackle the problems. They have measured the amounts of pollen produced by different kinds of plants. They have compared these amounts among species and between vegetation types. They have looked at the year-to-year variability in pollen production within and between populations of species. They have studied the types of pollinators that disperse the pollen grains. They have found that species with catkins or cones, such as oak and pine, produce tremendous amounts of pollen grains that are widely dispersed by the wind. In contrast, species with showy or fragrant flowers, like basswood and magnolias, produce relatively few pollen grains and rely on bees and other insects to carry pollen from one flower to another.

Researchers have also determined the relationships between pollen production and dispersal into lakes of different sizes and have related these figures to the abundance of plant species growing on local and regional areas beyond the lakes. Quaternary paleoecology is today a quantitative science, and Quaternary paleoecologists can be objective in interpreting the changes in abundance of pollen types through time as represented in fossil pollen records. So, rather than relying only on extensive field experience (which is also invaluable), there is an objective basis for understanding the ecological relationships of the fossil assemblages and for interpreting vegetation history. Confidence and credibility in landscape reconstructions are made possible not only because of statistically reliable data but also because relationships between forest composition around a lake site and the pollen assemblages settling on the surface of the lake mud can be determined.

As a basis for interpreting individual sites and making regional maps of vegetation history of the southeastern United States, in 1974 Paul and I began to build a database from modern pollen samples paired with forestry data from the Continuous Forest Inventory (CFI) of the United States Forest Service. First, we subdivided a map of the territory south of 36° N latitude and east of Texas and Oklahoma into 1° latitude by 1° longitude grid blocks. This size of map grid would give us several hundred samples collected on a spatial scale comparable with those being developed by Professor Thompson Webb, III, of Brown University and by Professor John McAndrews of the Royal Ontario Museum in Toronto. When the modern pollen database was finally combined in 1983, it would include some 1700 samples that spanned the representative life zones of eastern North America, from arctic tundra around Hudson Bay, to boreal forests of southern Canada, temperate forests of the Northeast and the Midwest regions of the United States, to grasslands of the Great Plains, and finally to warm-temperate and subtropical vegetation of the southeastern United States.

During the course of our graduate studies, we traveled across the southeastern United States, collecting pollen samples from lakes and wetlands. In the southeastern United States, there are two districts in which natural lakes are abundant. One of those is along the Atlantic Coastal Plain from the Delmarva Peninsula to South Carolina. On the Atlantic Coastal Plain, hundreds of thousands of elliptical basins, known as Carolina Bays, dot the landscape. Most of the basins are filled in with peat, and almost all of those have been converted to truck farms. Notable exceptions include open-water lakes such as Singletary Lake in Bladen County, North Carolina. People have long speculated on the origin of the Carolina Bay lakes. They are generally oriented with their elliptical long axis extending from northwest to southeast, and have a sandy rim at their southeastern side. Some investigators have claimed that Carolina Bays originated as alligator or even mastodon wallows. Others have called for a cosmic event, such as a meteor shower, to have carved out the many lakes. Geologists have determined several more plausible causes — bay lakes to the north, from Delaware to Virginia and North Carolina, probably originated as thaw lakes in Pleistocene permafrost, oriented by prevailing wind as those on the north slope of Alaska form today. Those to the south, on the coastal plain of South Carolina and Georgia, may have originated as blowouts on ancient sand dunes, with their orientation determined by the prevailing westerly winds during the Pleistocene.

The second lake district in the southeastern United States is along the panhandle and peninsula of Florida. The state of Florida lies largely on a limestone platform, formed as ancient coral reefs in a higher Tertiary sea. During the times of lowered sea level associated with glacial intervals, Florida lakes formed in sinkholes within the limestone, and are most abundant along the highland spine that runs north to south through the center of the Florida peninsula.

From the southern Appalachian Mountains to the Ozark highlands and south to the Gulf of Mexico, naturally occurring lakes are scarce. Sinkholes dissolved into limestone or dolomite bedrock occur in central Kentucky and Tennessee and on the plateaus of the Ozark highlands. Many of those sinkholes, however, hold water only during the winter and dry out seasonally, a situation in which pollen grains are poorly preserved. In collecting modern pollen samples, as well as in finding pollen sites of late Pleistocene age, we had to develop a geological search strategy that would help us select potentially productive sites from which to collect our pollen samples.

In collecting modern pollen samples, our first priority was to locate natural lakes, swamps, and marshes. If we could not find such a site in a given grid square on our map, we looked for artificial lakes or ponds — here, we were only interested in samples of surface mud, so reservoirs or fishing ponds would give us a sample of the modern pollen rain. We even considered water traps on golf courses as potentially suitable sites. If the area was so well drained that there were no bodies of water, we collected patches of moss from clearings in the forest. The clumps, or polsters, of moss are good collectors of pollen grains, because they tend to stay moist, they are usually acidic, and they can preserve well the pollen grains that accumulate over a five- to ten-year interval.

From our grid of modern pollen samples, we mapped out the changes in importance of different pollen types across a wide geographic region. We found that oak and hickory pollen are most abundant in the mid-south region from the southern Appalachian Mountains to the Ozark highlands. Bald cypress pollen is characteristically found in swamplands along the Mississippi and Atchafalaya rivers of Louisiana. Pine pollen dominates the modern pollen assemblage across the Gulf and southern Atlantic coastal plains. In order to use this information in reading landscapes of the past, we needed to know how the changes in pollen across the region corresponded with changes in the composition of the forest vegetation. If we could calibrate pollen to vegetation, then we could use the calibrated relationships to better interpret forest history.

As a basis for comparison, we went to the field records of the foresters who tally the number of board-feet of wood commercially available in tracts of merchantable timber. These survey crews, employed by the United States Forest Service, search the region, county by county, state by state, completing one circuit and publishing summary documents every ten years. This process and the summary documents are called the Continuous Forest Inventory (CFI).

CFI records are available as published lists of the two dozen or so most important tree species in each county. The measure most useful to ecologists is growing stock volume, that is, the wood volume of merchantable wood in the trunks of the trees of each species. This is the healthy, growing stock of the tree that can be harvested and cut into boards. Growing stock volume also gives a good estimate of the ecological importance of the species in the forest because it reflects the extent of canopy cover of the tree and hence the amount of area the tree dominates in the forest stand. Canopy cover in turn is directly related to the amount of pollen produced by a species in relation to other trees in a stand and hence to the potential contribution of a tree species in the pollen record.

From the CFI records, we developed a computer database to compare pollen percentages from surface samples with tree percentages, county by county across the southeastern United States. In order to visualize the relationships of pollen to vegetation across real landscapes, we plotted percentage values on a regional map. Then we compared the relationships statistically in order to develop equations for converting fossil pollen percentages into estimates of abundance of trees in today's forests. We hoped to apply the modern relationship between forest composition and pollen representation to fossil pollen assemblages in order to interpret forest history well back into the past.

It soon became clear that some tree species, for example, the many species of oak, produce pollen approximately in proportion to their abundance in the vegetation. Many tree species, however, produce a disproportionate amount of pollen, and their populations are thus either over-represented by their pollen percentages in lake sediments or else they are under-represented by their pollen. Pine trees, for example, produce abundant pollen grains that are readily dispersed by wind, and they may contribute more than ten percent of the pollen assemblage in a sample of lake mud without pine trees actually being present on the local watershed. On the other hand, populations of trees such as maple and tulip tree that are mainly insect-pollinated and that do not produce pollen grains abundantly are greatly under-represented in the pollen record. These species may make up as much as twenty percent of a forest stand and still contribute only one or a few percent of the total pollen rain to a nearby lake surface. This disparity in pollen representation may be accentuated if dominant trees on the watershed are also prolific pollen producers, effectively diluting the contribution of those species that are inherently low pollen producers and making them even less visible on the landscapes of the past.

Most tree species, however, lie somewhere in between these two extremes of either exceptionally abundant or very poor pollen production. Knowing the general characteristics of the production and dispersal of pollen makes interpreting their former abundance on past landscapes more tractable. No longer can we assume, for instance, that because pine was once forty percent of the fossil pollen assemblage, it must also have been forty percent of the trees. We can say, however, that pine was probably somewhat less important than forty percent of the vegetation, and that tulip tree, whose pollen occurred only in trace amounts, nevertheless was probably far more important than its small fraction in the pollen record implies at first glance. In the process of comparing CFI records with modern pollen samples, we began to learn how to more accurately translate the language of pollen into vegetation. Thus we came a step closer to understanding the story that pollen tells and to visualizing landscapes of the past based upon fossil pollen assemblages from southeastern lakes, bogs, and stream terraces. This helped to prepare us for the challenge of locating Pleistocene refuges for deciduous forest species and for mapping the Holocene emergence of the eastern deciduous forest.

Part III

Looking for a Pleistocene Deciduous Forest

Testing the Blufflands Hypothesis

Our field work in the Tunica Hills had confirmed the previous findings of Clair Brown and supported the interpretation that the Tunica Hills were a haven for both hardwoods and spruce trees during the Pleistocene. Even so, upon leaving LSU for the University of Minnesota, we were acutely aware that the history of Pleistocene landscapes in the southeastern United States still remained largely unstudied.

In the 1970s, the vast, generally lakeless terrain west and south of the Appalachian Mountains remained uncharted by Quaternary paleoecologists. Yet, therein potentially lay the answers to some of the central questions that had been posed by generations of biogeographers. The crux of the problem was to find the location of the deciduous forest during the Pleistocene. Just how far was it displaced southward, and therefore what was the nature and extent of climate change in the southeastern United States? What were the former boundaries of the eastern deciduous forest, and what were the differences in abundance of tree species across their ranges during the late Pleistocene? Were the plant communities of the past the same as those of today, and if not, how did they differ? In what sequence did tree species spread northward to eventually blanket much of the formerly glaciated terrain north of the confluence of the Ohio and Mississippi rivers?

These questions could not be answered with evidence from any one site. Paleoecological studies from several as-yet-unstudied areas in the southeastern United States were required in order to unlock a series of windows onto the past that were gateways to our understanding of larger questions concerning the nature of world-wide climate change and the organization of biological communities.

Other Quaternary paleoecologists who had preceded us in the quest for finding a Pleistocene deciduous forest south of the glacial margin had found tantalizing, though largely negative, results. But, just as the

117

modern deciduous forest can be defined not only by what it is today, but also by *what it is not*, so too, we thought, could the whereabouts of Pleistocene deciduous forest be narrowed down by knowing *where it was not*. As students, we needed to learn what was already known, then develop plans for our research that would take us beyond the bounds of previous knowledge. We needed to frame testable hypotheses and then develop an extended and systematic geobotanical search strategy to locate just the right sites to answer the myriad questions about the history of the eastern deciduous forest — all within a geographic region in which other Quaternary paleoecologists, including our mentor, Herb Wright, thought no suitable field sites existed.

Across the vast region from the Ozark highlands to the central Atlantic Coastal Plain, only a few well-dated sites had been studied prior to the 1970s. On the western margin of the modern eastern deciduous forest, in western Missouri, stream-terrace deposits such as Boney Spring along the Pomme de Terre River had yielded evidence of Pleistocene spruce forest. Studied in the 1960s by Professor Peter Mehringer of Washington State University and by Jim King from the Illinois State Museum, some of these spring deposits dated from the last full-glacial interval but they contained no evidence of temperate deciduous forest. Mehringer and King's studies refuted the previous notion of Steyermark that the Ozark highlands were an ancient center of evolutionary origin for species of oak and hickory and that the Ozark plateaus were a stable refuge unaffected by climate change during the Pleistocene. The records from terraces along the Pomme de Terre River, however, contained no plant fossils from late-glacial or Holocene sediments to indicate when oak-hickory forest had become established in the Ozark highlands. This absence of evidence left many questions unanswered about both the location of refuges and the subsequent emergence of deciduous forest communities.

Pittsburg Basin, a site located just beyond the glacial margin in southern Illinois, posed a major challenge to paleoecological interpretation. This site was first studied by Eberhard Grüger, a German palynologist, under the sponsorship of Professor Wright and the University of Minnesota Limnological Research Center (known to its alumni and associates as the LRC). Pittsburg Basin was a small kettle lake formed on Illinoian glacial till. It existed during the interglacial interval that preceded the present one, as well as throughout the most recent, or Wisconsinan, glaciation, during which it lay beyond the margin of the Laurentide Ice Sheet. The pollen record from Pittsburg Basin proved enigmatic. It contained large percentages of deciduous tree pollen throughout the Wisconsinan interval, in addition to pollen of many herbs. It was only upon re-examination of this record and measuring the absolute amounts of pollen of each pollen type rather than the relative proportions that the meaning of the site became clear. Jim King, who recored and re-analyzed the Pittsburg Basin site, discovered that the total amounts of pollen in the lake sediments were very low, comparable with treeless environments inhabited by plants with low pollen production, not a forested landscape that would produce a much greater pollen rain. Rather than representing late Pleistocene deciduous forest, the Pittsburg Basin pollen record reflected a treeless tundra environment into which a few grains of oak pollen had evidently drifted from somewhere far to the south.

South of the glacial margin across Kentucky and Tennessee, the map of Pleistocene vegetation was blank in the early 1970s. Farther to the east, in the central Appalachian Mountains, high-elevation bogs were studied by Paul Martin (Marsh, Pennsylvania) and by Margaret Davis and her student Jean Maxwell (Buckles Bog, Maryland). In these sites, high percentages of herb pollen and low absolute pollen amounts confirmed that much of the Appalachian Mountains served as a Pleistocene refuge, not for deciduous forest, but for arctic-alpine tundra. Tundra communities grew as far south as Cranberry Glades on the Allegheny Plateau of West Virginia. Cranberry Glades was studied by Professor William A. Watts of Trinity College, Dublin, Ireland, who worked extensively in eastern North America from the 1960s through the 1990s under the auspices of the LRC. The work of Martin, Davis, and Watts in the central and southern Appalachian Mountains served as a counterpoint to suppositions made by Stanley Cain about the antiquity of the mixed mesophytic forest. If the summits of the southern Appalachian Mountains harbored arctic-alpine tundra during the Pleistocene, then it was clear that climate change had pervasive effects throughout the region that today is occupied by the eastern deciduous forest.

The Carolina Bays of North Carolina were initially studied in the 1940s by the husband and wife team of Murray and Helen Buell from Rutgers University. Their early studies of Jerome Bay were followed up in the 1950s and 1960s by professors David Frey and Donald White-head of Indiana University. Frey and Whitehead studied late Pleistocene sediments from Singletary Lake and found overwhelming evidence of pine forest with only a small amount of spruce and no pollen of temperate deciduous trees. This discovery led to a study of the distinguishing characteristics of pine pollen in order to find out whether northern or southern species of pine composed the glacial-age forests of the Atlantic Coastal Plain. The size of the full-glacial pine pollen from Singletary Lake matched that of two northern species, jack pine and red pine, and thus confirmed the interpretation that a cold climate had existed in the region during the late Pleistocene.

Beginning in the 1960s, Bill Watts embarked upon a quest to find the Pleistocene deciduous forest. Watts was driven by his interest in the Arcto-Tertiary Geoflora, which had largely vanished from the British Isles and continental Europe by the beginning of the Pleistocene. Plants he knew from Tertiary-age deposits in Europe are, however, still living in the southeastern United States. The history of temperate deciduous forest was therefore not only a major curiosity to Watts, but also it took on a special significance for him because of the contrasting Quaternary histories of Europe and eastern North America.

Bill Watts began his search for Pleistocene deciduous forest on the Piedmont of northwestern Georgia. At Bob Black and Quicksand ponds, located in Bartow County, Georgia, he found evidence of major changes in climate and vegetation through the past twenty thousand years. During the Wisconsinan full-glacial interval, jack pine dominated the forests of northern Georgia, as evidenced by both pollen and needle-like leaves whose cell structures matched those of the boreal species *Pinus banksiana*. Deciduous forest replaced boreal-like conifer forest in northern Georgia after about sixteen thousand radiocarbon years ago, with pollen of oak, ash, hornbeam, and other temperate trees increasing significantly throughout the late-glacial and postglacial intervals.

Even as far south as South Carolina, Watts found that Carolina

Bay lakes such as White Pond, located on the inner coastal plain near Columbia, contained fossil pollen primarily of jack pine and spruce, with little evidence of temperate deciduous trees. At White Pond, rich, mixed deciduous forest in which American beech was a dominant species arose only late in the late-glacial interval. Evidently American beech, hickory, and other temperate deciduous trees migrated onto the coastal plain of the Carolinas after climate warming made those environments inhospitable for boreal conifers.

Watts and Wright made many field trips to Florida in search of Pleistocene pollen records. The Florida lakes they studied, however, provided little added information on the whereabouts of deciduous forest species during the Wisconsinan glacial maximum. Many of the sinkhole lakes in Florida contained sediments dating only from the Holocene interglacial. They apparently were dry basins during the late Pleistocene. At the southernmost tip of the central highland spine of Florida, one site yielded spectacular although unexpected results. Lake Annie proved to be the oldest lake site found in Florida by Watts and Wright. Located on the grounds of the Archbold Biological Station, Lake Annie is almost twenty meters deep and contains another twenty meters of pollen-rich sediment beneath the lake bottom. The fossil record from Lake Annie is estimated to date back more than one hundred thousand years. The pollen assemblages show wide swings in composition through the record, but very sharp changes in the proportions of pine and oak pollen occur just in the sediments that ought to record events during the last, or Wisconsinan, glacial maximum. Many attempts at radiocarbon dating have shown that there is a gap, or a hiatus, in the lake sediments during that interval of time. Watts and Wright concluded that Lake Annie was dry during a very significant portion of the late Pleistocene. Pollen assemblages from just before and just after the hiatus contain high proportions of sand dune scrub plants such as the native rosemary (*Ceratiola ericoides*). Bill Watts's work has shown that southern Florida was a very inhospitable place for mixed mesophytic forest species during the past one hundred thousand years and thus was ruled out as a location for Deevey's late-Pleistocene deciduous forest.

Watts also was drawn to Mexico by the earlier work of Paul Sears and Katherine Clisby. From a site in the central Valley of Mexico called Lake Patzquaro, Watts studied a long sediment sequence that spanned the late Pleistocene and Holocene. The pollen record from Lake Patzquaro showed very little change during the time at which glaciers reached their maximum extent at high latitudes during the late Wisconsinan glaciation. The original finds of spruce pollen, reported by Sears and Clisby, could not be duplicated at Lake Patzquaro. Instead, the "complacent" pollen record from that lake was cited as evidence that neither boreal nor cool-temperate species had been displaced as far southward as predicted by Deevey in his 1949 monograph.

In the 1950s, Paul Martin had become interested in another biogeographic puzzle related to testing Deevey's speculation that deciduous forest species had migrated far to the south under the stress of Pleistocene climate cooling. Martin studied the distributions of plethodontid (lungless) salamanders (Plethedontidae) in the southeastern United States and Mexico. He found that, as had been reported for vascular plants by Jack Sharp, a number of species of plethodontids occurred in both the southern Appalachian Mountains and the cloud forests of the Sierra Madre Orientale of eastern Mexico. Martin proposed a "Texas corridor" for the interchange of species and speculated that this migration corridor might have been used by dispersing plethodontids as recently as the Pleistocene. The alternative to Martin's model was that a separation had occurred in the species' ranges that was much more ancient. This second model required that populations of plethodontid salamanders, along with species of plants that included white pine, sweetgum, and partridge berry (*Mitchella repens*), had remained disjunct in the two mountain ranges since the late Tertiary.

Vaughan Bryant and Richard Holloway tested the Texas corridor hypothesis in the 1970s through a restudy of ponds in southeastern Texas. Earlier, Potzger and Tharp had reported spruce pollen in the deepest sediments of several spring-fed ponds aligned along the Carizzo Aquifer of southeastern Texas. Upon reinvestigation, Bryant and Holloway did find several percent of spruce pollen, dated to at least sixteen thousand radiocarbon years ago at Boriack Bog, confirming that substantial cooling of climates had occurred there during the last glacial maximum. With postglacial warming of climates, however, spruce forest changed to dry woodland with no trace of a species-rich, mesic forest that could have supported either the species of plants or those of salamanders with the classic disjunct distributions. Bryant and Holloway showed convincingly that during the late Pleistocene, no extensive Texas corridor was available for migrations either of plethodontid salamanders or of mixed mesophytic forest species — it had been too dry. Instead, the separation in ranges must have occurred much earlier in the Pleistocene or even before in the Tertiary.

By the time we were ready to begin our research toward Ph.D. degrees at the LRC in 1974, studies such as those of Watts and of Bryant and Holloway had shown clearly that neither southern Florida nor central Mexico served as a refuge for the Pleistocene deciduous forest. Neither the Ozark highlands nor the southern Appalachian Mountains supported deciduous forest during the late Pleistocene, and the Atlantic Coastal Plain was also largely ruled out as a significant refuge area. Paul and I thought that the answer to the conundrum posed by Braun and Deevey must lie somewhere in between those geographic areas that already had been excluded as potential refuges for Pleistocene deciduous forest.

For our dissertations, therefore, we embarked on developing a research strategy expanded from the studies we had begun while at LSU. We first needed to test our hypothesis that the Blufflands had served as a migration pathway for spruce and deciduous forest species during the late Pleistocene and Holocene. To accomplish our objective, we needed to bolster the evidence from the Tunica Hills with fossils collected from a second site in the Blufflands that included a mixture of spruce and hardwoods during the last full-glacial interval. We also needed to study sites east of the central Mississippi Valley to document the Pleistocene vegetation away from the Blufflands. A significant portion of that research would require finding a full-glacial pollen site within the heartland of Braun's mixed mesophytic forest. Finally, if studies in the mixed mesophytic forest were to confirm that boreal conifer forests ruled the mid-latitudes of Tennessee during the late Pleistocene, then we would also need to search farther south onto the Gulf Coastal Plain for evidence of deciduous forest refuges. Over the course of two field seasons, 1975-1976, we canvassed the region west of the southern Appalachian Mountains to the Mississippi Valley and from Tennessee south to the Gulf Coastal Plain looking for fossil sites with sediment records that would extend back to at least twenty thousand radiocarbon years ago. We found several sites on those expeditions that became the basis of our dissertations. I concentrated on two ponds on the eastern Highland Rim of Tennessee. Paul studied two contrasting sites, one located in the Blufflands of western Tennessee, and the other on the Gulf Coastal Plain of southern Alabama. Taken together, data from these fossil sites and from additional sites studied by our predecessors and our students in the surrounding region have allowed us to piece together the late Pleistocene and Holocene history of vegetation and climate south of the glacial margin in eastern North America.

The key to unlocking one of the most important of these windows into the past was found in 1976. That spring, two brothers were playing on the bank of a creek channel near the interchange of I-240 and Perkins Road in downtown Memphis, Tennessee, near the construction site of the Greater Memphis Mall. One of the boys noticed an ivory tusk sticking out of the stream bank. Their father contacted Ronald Brister, Curator of Collections at the Memphis Pink Palace Museum, and soon thereafter an excavation of the Nonconnah Creek mastodon was underway. We were primed for the opportunity of studying a full-glacial site in the Blufflands of western Tennessee, and when we were informed of this prospect Paul and I drove immediately to Memphis to help excavate the Nonconnah Creek mastodon site. What we found there became a significant portion of Paul's dissertation.

The Nonconnah Creek site was completely exposed above the water table. Most of the mastodon's skeleton had been destroyed when the creek channel was deepened by dredging and straightened by heavy earth-moving equipment. But the context of the mastodon was still there. The skull was buried in a thick layer of organic-rich, black clay. The clay lens was more than a meter thick. Beneath the clay layer was a sand and cobble stream bed, and above it was a cap of wind-blown, silty loess.

After our work with stream terraces in the Tunica Hills of Louisiana, we were familiar with the geologic setting that Nonconnah Creek posed. To study the plant-fossil record of organic lenses exposed above the water line in stream terrace deposits, we needed to use a sampling technique very different than that used to core the sediment of a pond or lake from a floating raft with the square-rod piston sampler (Figure 20). At Nonconnah Creek, first we cleaned off the face of the cut bank with shovels and sharpened trowels. Then we carefully described the stratigraphic layering of the sediments in which the mastodon skull was embedded. We collected samples of the organic-rich, black clay by cutting contiguous blocks out of the vertical cliff face. Each block of sediment measured ten centimeters wide, ten centimeters deep, and five

Figure 20. Hazel Delcourt at full-glacial locality for mastodon, spruce, and temperate deciduous trees at Nonconnah Creek, Memphis, Tennessee. (Photograph by Paul A. Delcourt)

125

centimeters high. We kept each sediment block right side up and carefully labeled and wrapped each one. In this way we took several "monoliths," or vertical columns of sediment. One set of samples was available to sieve for plant macrofossils, one set would be used to sample and process for fossil pollen analysis and radiocarbon dating, and one set was kept in the Pink Palace Museum for future reference.

Almost immediately upon cutting into the fossil-rich earth next to the mastodon skull, we found exciting evidence of the late Pleistocene environment. We recovered both small and large spruce cones — cones of the modern species of black spruce (*Picea mariana*) as well as of the now-extinct *Picea critchfieldii*. The largest of the spruce cones resembled those of modern white spruce, except, as had been the case in the Tunica Hills, they were much larger. Some of the Nonconnah Creek spruce cones were fully ten centimeters long, far outside the size range of any living species of eastern North American *Picea*. We also found many spruce twigs. These are easy to identify even in the field because of their short, stubby side shoots that form pegs to which the needles were once attached.

What we found next astounded us. It was the carbonized remains of a walnut shell, crushed between the mastodon's molars. We dated the mastodon as more than seventeen thousand radiocarbon years old based upon the walnut and several spruce cones we collected from within the mastodon's mouth, its last dinner. The late Pleistocene assemblage of fossil plants included not only spruce and walnut, but a number of additional cool-temperate deciduous forest species. We found hickory nuts, an acorn, winged tulip tree seeds, and pollen grains of hazelnut, sugar maple, and American beech in the glacial-age assemblage of plant fossils.

The Nonconnah Creek locality was an exquisite prize. It gave us an important piece of the biogeographic jigsaw puzzle we sought to piece together with our dissertation studies. Nonconnah Creek is located in the northern Blufflands and it provided a direct test of our Blufflands migration corridor hypothesis. From this fossil site, it became clear that not only the Tunica Hills, but also the northern Blufflands, were a haven for spruce and hardwoods during the late Pleistocene (Figure 21). Whether or not this intermingled community

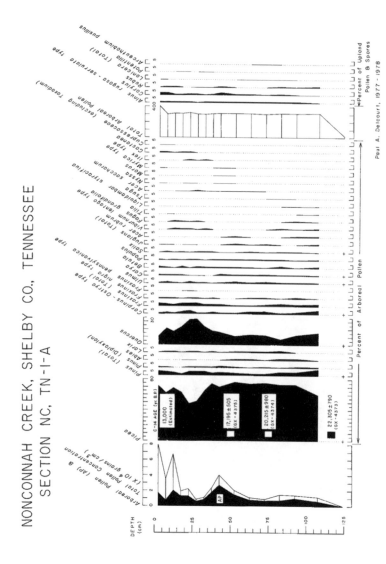

Figure 21. Pollen diagram for the full-glacial locality at Nonconnah Creek, Memphis, Tennessee. (Diagram from Delcourt and Delcourt, 1996)

of boreal and cool-temperate plants was more widespread beyond the narrow Blufflands corridor was a question that we investigated with another fossil site located in the heartland of E. Lucy Braun's beloved mixed mesophytic forest.

Pleistocene Boreal Forest in the Mixed Mesophytic Forest Heartland

For my dissertation, I set out to test one of Braun's central assumptions. She had envisioned that the Allegheny and Cumberland plateaus were ancient and stable landforms that had served as refuges for her rich and diverse mixed mesophytic forest, the centerpiece of the eastern deciduous forest. Even though a number of late Pleistocene sites had been studied around the periphery of the mixed mesophytic forest, Braun's assertion was still controversial and would remain so until someone tackled it head-on, and I decided to do just that.

I reasoned that if temperate deciduous forest species had survived the last full-glacial interval in the southern Appalachian Mountain region, one of the most likely areas for this to have taken place would have been on or near the dissected southern part of the Cumberland Plateau in Tennessee. This plateau region is lower in elevation than are most of the southern Blue Ridge Mountains to the east. The upland surface of the Cumberland Plateau is generally flat-lying and is capped by sandstone that is relatively resistant to weathering. Where eroded by streams, however, the sandstone cap forms high cliffs above deeply dissected gorges (Figure 22). Today, the vegetation on the surface of the southern Cumberland Plateau is mainly oak forest, but mixed mesophytic forest communities are found on moist mid-slopes, where they grow on calcium-rich soils weathered out of the limestone and the dolomite from geologic formations that lie underneath the sandstone caprock. In the past, the rugged terrain in and adjacent to the Cumberland Plateau may have afforded an appropriate microclimate for mixed mesophytic forest species to have survived throughout the Pleistocene.

The sandstone layer that tops the Cumberland Plateau is gener-

Figure 22. Frayed western edge of the Cumberland Plateau in middle Tennessee. Local relief is three hundred meters. (Photograph by Hazel R. Delcourt)

ally too well drained for the formation of permanent ponds and swamps that might give evidence of changes in vegetation and climate. We knew of a few small marshes at the headwaters of streams that drained the plateau, but those wetlands were very shallow, with poorly preserved fossils. It occurred to us that a second strategy for finding an appropriate site to test Braun's hypothesis would be to look for deeper sinkholes formed by solution of limestone bedrock and located at the foot of the western slopes of the Cumberland Plateau. In the LRC library, we searched the terrain portrayed on USGS topographic maps along the plateau's frayed western edge until we found a quadrangle called "Dry Valley" that contained a number of what appeared to be permanently wet sinkholes. The sinkholes were located within a kilometer or less distance from steep slopes and coves that today harbor mixed meso-phytic forest communities — a location within the pollen dispersal range of many of the types of deciduous trees that grow along the edge of the plateau. With the hope that we would find among these sinkholes a site that had formed during the last full-glacial interval, during the summer field season of 1975 we set out with a trailer full of coring equipment on our first trip into the heartland of the mixed mesophytic forest region.

Traveling south from Cookeville to Sparta, Tennessee, I noted the gently rolling terrain of the Eastern Highland Rim, the broad bench of land between the Nashville Basin to the west and the Cumberland Plateau to the east. Pine and oak forest was barely visible on the crest of the Cumberland Plateau. Downslope, the forest changed to mixed hardwoods that filled the steep gorges along the western margin of the plateau with their richly textured greens. In the valleys and on the Eastern Highland Rim, tobacco (*Nicotiana tabacum*) fields and tree nurseries replaced forest. On my printed map of the Dry Valley topographic quadrangle, numerous sinkholes dotted the level to rolling landscape of the Eastern Highland Rim. I knew that the land surface was underlain by Mississippian-age limestone and dolomite. Although most limestone sinks in the Nashville Basin drained into underground streams, I was hoping that on the Eastern Highland Rim I would find the illusive prize — my window into the past, in the form of a deep and permanent sinkhole lake near Sparta. I was somewhat dismayed to realize that the topographic map I held must have been based on aerial photos shot in winter, when a surplus of rain and low evaporation left much of the terrain inundated. In late summer, when we were looking for permanently wet sinkholes, most of them were dry with only a thicket of tag alder (*Alnus rugosa*) and red maple to mark the locations of winter wetlands.

A large sinkhole, fed by several incoming streams and surrounded by tobacco warehouses and middle-class subdivisions, became our primary destination. This sinkhole was likely the only permanent body of standing water in the county. At the edge of the wetland, a white sign with enormous red letters proclaimed how undesirable a landscape feature an active sinkhole is to most Tennesseans. The sign read "Dump Fill Dirt Here."

After finding a place to park near the edge of the swamp, I peered cautiously through the star-shaped foliage of a sweetgum tree into its interior. From my vantage point, I saw enough to confirm that it was, indeed, very wet, and very muddy beyond the forest edge. Before we invested time and energy assembling the coring device, I needed assurance that the pond was old, for if it had formed only in the past few decades, sediments on its bottom would not contain the secrets of the ancient past that I sought. Bushwhacking through the vegetation, Paul and I stumbled upon a clue — a gravestone dating back to the 1700s. On it were faint letters spelling "Anderson." In my field notes, I named this site Anderson Pond. Because a permanent pioneer settlement had been located on the rim of the pond, it probably had been a reliable source of drinking water even two hundred years ago. I surmised that this sinkhole had not collapsed recently, and that it was probably quite ancient. If so, and if in most years it had stayed wet year-around, then the mud covering the lake bottom might be quite thick and might contain abundant fossil remains of former plant life. Behind his thick-rimmed glasses, Paul's blue-green eyes sparkled with the anticipation of discovery. This place felt right to him, too.

After having searched for viable wetland sites east of Memphis for the previous week to no avail, I was more than ready to get my feet wet. With the landowners' permission, it was time for the ultimate test — time to probe the depths of the swamp. Slowly I waded into the dense thicket of shrubs at the edge of the wetland. Sticky clay clung fast to my feet, threatening to suck off my leather boots. Paul and I needed maneuvering room for the coring device. We needed to find an area of open water, free of tree roots and aquatic plants. Based on aerial photographs I had studied earlier, I thought we should be able to find a natural clearing in the interior of the swamp. Paul slowly cleared a path with his machete. My senses became filled with the rotten-egg odor of swamp gas as methane produced by the decaying organic matter began to rise from beneath our feet.

I had plenty of time to botanize on the way to the center of Anderson Pond (Figure 23). The swamp thicket was composed of Virginia willow (*Itea virginica*), a shoulder-high shrub with oval leaves, and swamp loosestrife (*Decodon verticillatus*), a willowy, pink-flowered shrub that forms an almost impenetrable tangle of slender, pliable stems. Scattered through the wetland were old stumps on which red maples and alders had taken root. Their root crowns formed occasional stepping stones and were a welcome reprieve for water-logged feet as we hop-scotched from knoll to knoll. The expedition was becoming a lei-

Figure 23. Swamp thicket and pools in the interior of Anderson Pond. (Photograph by Hazel R. Delcourt)

surely morning of bushwhacking through the Anderson Pond swamp, as we were searching for an extensive pond of open water that, as it turned out, did not exist. Rather, after climbing a red maple tree and ascertaining that we were about in the center of the basin, Paul declared that it was time to try the sediment probe.

The first drive of the corer came easily. It penetrated downward some three feet into the muck through which we had been slogging knee-deep. The mud was dark gray and had the consistency of modeling clay. A second drive of the corer extended farther beneath the first. We found more gray clay, but this time it was densely compacted. With only the two of us to hold the piston and push the sampler into the mud, we needed mechanical assistance to probe any farther into the stiff silty clay. We screwed two earth anchors, the same kind of large screws with handles that are commonly used to brace telephone poles, into the ground about two meters apart. Then we attached a T-shaped yoke to the top of the coring rod and attached a stout rope to one earth anchor, strung it through eyelets attached to the yoke, and connected it to the other earth anchor with a come-along. The come-along, also called a chain hoist, had chains on either end. By racheting the chain hoist, we could shorten the chain, which used its mechanical advantage to tighten the rope and drove the coring device into the stiff sediment. Slowly we took one more drive with the coring device, then another, and another. To our delight, we were able to probe deeper than we had dared to hope, about seven meters in all.

Toward the end of the day, with six long core segments stowed away safely in the back of the Land Cruiser and the seventh still in the depths of the pond sediments, a sudden thunderstorm threatened to end our days as paleoecologists prematurely. As the pond began to swell with runoff from flashy inflowing streams, we struggled to extricate the core barrel from the sticky clay. As we clambered up the bank at the pond's edge, a bolt of lightning struck nearby. It wasn't until the next day that Paul told me he had felt the electrical current run through the core barrel into his arms. It was a close call, but we saved both the sediment core and Wright's coring device.

Thus, our first coring expedition to Anderson Pond was successful. Once back at the LRC, I prepared a series of sediment samples for pollen analysis. I broke the suspense by looking at the lowermost sediment samples first. They were composed of more than eighty percent pine pollen, and the remainder was largely spruce pollen. There were only traces of pollen of oak and other deciduous trees. That pollen assemblage was characteristic of the lower four and one-half meters of sediment. Only in the upper one and one-half meters did pollen percentages of temperate deciduous trees increase. The Anderson Pond site not only recorded vegetation history since the late Pleistocene, but its pollen record was strikingly similar to those studied by Watts from Bob Black and Quicksand ponds in northern Georgia. What I had found in the sediments of Anderson Pond was evidence of a late Pleistocene forest unlike any found in the state of Tennessee today (Figure 24).

In 1976 we returned to Anderson Pond with fellow student Steve Lund to probe in a series of cores across the sinkhole basin, with the aim of obtaining the most complete set of sediment cores possible for detailed radiocarbon dating, pollen and plant macrofossil analysis, and sedimentologic analysis that included paleomagnetic stratigraphy. From the paleomagnetic measurements, which required that the sediment cores were taken knowing their exact orientation with respect to magnetic north, Steve would determine changes in source area of sediment and changes in Earth's magnetic field that reflected changes in climate through the history of the lake site.

Beneath the surface sediments, the Anderson Pond basin is shaped like a petri dish, steep-sided and flat-bottomed. Underneath the general floor of the sinkhole are local chutes weathered into the underlying limestone bedrock before the collapse of the sinkhole basin. As we probed across the basin, we cored a full three meters into one of those chutes, which was filled with red soil and weathered limestone rubble. A lens of organic matter recovered from ten meters depth dated to twenty-five thousand radiocarbon years ago. Similar crevices weathered through limestone bedrock are visible in road cuts along Interstate 40 near Cookeville, Tennessee.

Radiocarbon dates confirmed that the organic-rich, gray silty clay that filled the Anderson Pond basin to a depth of seven meters began accumulating some nineteen thousand radiocarbon years ago, during the height of the last glacial maximum. Because these lake sediments

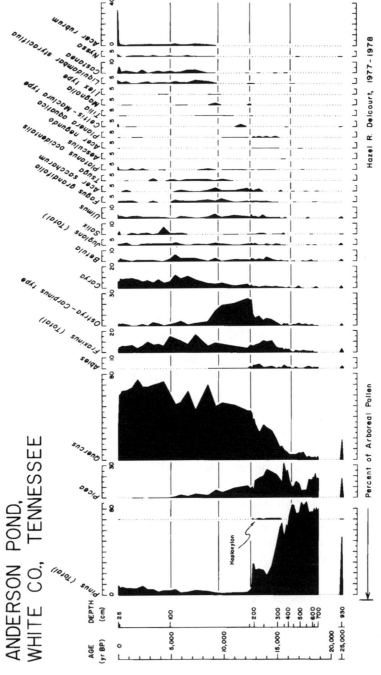

ANDERSON POND, WHITE CO., TENNESSEE

Figure 24. Pollen diagram for Anderson Pond, White County, Tennessee. (Diagram from Delcourt, 1979)

were uniform in both color and texture from bottom to top, we needed a series of radiocarbon dates to tell us whether rates of sediment accumulation had been uniform through time or whether they varied widely. The dates confirmed what the preliminary analysis of pollen assemblages had indicated about the history of the site. Only the uppermost meter of sediment dated from the last ten thousand years — and the lowermost six meters of silty clay had accumulated over just the previous nine thousand years. There was an abrupt change in rate of sediment accumulation about twelve thousand five hundred radiocarbon years ago that corresponded with a decrease in source area for sediment as determined from the clay minerals. During the late Pleistocene, sediment washed into the pond from as much as a kilometer away on the Cumberland Plateau, whereas during the Holocene, the sediment source was from the local watershed. After the collapse of the petri dish-shaped Anderson Pond basin, clay apparently settled on its floor and formed a seal that perched the water table. During the late Pleistocene, the sinkhole basin became a seven-meter deep, permanent pond that thereafter gradually filled in with fine clayey mud and plant debris.

From looking at the fossils of aquatic plants in the sediments, I was able to further interpret the changes in sediment accumulation rates as changes in aquatic environment through time (Figure 24). Aquatic plants living in Anderson Pond during the late Pleistocene are preserved as both fossil pollen and as fruit and seed macrofossils, including both submersed and floating forms of pondweed (*Potamogeton*) and naiad (*Najas*) that today live in small lakes within the upper Great Lakes region of Minnesota, Wisconsin, Michigan, and Ontario. They require cold, clear water and stable year-around water levels. Through the late Pleistocene, nearly a centimeter of silt and clay accumulated on the bottom of Anderson Pond every hundred years. As the lake filled in it became shallower, and rooted aquatic plants such as water lilies became established. At the transition from late Pleistocene to Holocene, the pond was shallow enough to have a marsh of sedge and grass around its perimeter. Shrubs including buttonbush (*Cephalanthus occidentalis*), tag alder, and red maple established in the swamp in the early Holocene, when the pond was shallow and when the climate had changed from cool conditions year-around to alternating cool, wet winters and warm, drier summers. The history of changes in the aquatic environment is mirrored by changes in diatom assemblages. As studied by Robert Tolliver for his dissertation, these single-celled algae proved to be excellent indicators of the changeover from a deep, clear-water pond to a shallow shrub-swamp at the Pleistocene-Holocene transition.

The full-glacial pollen spectra from Anderson Pond, dating from seventeen to nineteen thousand radiocarbon years before present, were overwhelmingly dominated by northern pine pollen (Figure 24). From measurements of the fossil pine pollen grains, I determined that either jack pine or red pine was the species represented. Fragments of pine needles that I screened from the clay matched those of jack pine.

The closest modern analog for the Pleistocene jack pine forest of Tennessee is located near Winnepeg in central Manitoba, which today is more than eighty percent jack pine forest. Even accounting for overrepresentation of jack pine by its pollen, the full-glacial landscape surrounding Anderson Pond must have been mostly boreal-like pine forest. The very few grains of oak, ash, and hornbeam pollen in full-glacial sediments from Anderson Pond may have come from trees growing locally in valleys of the Cumberland Plateau. Alternatively, those pollen grains may have blown in from deciduous forests growing at some distance to the south, possibly as far away as Alabama.

At Anderson Pond, we had visited the heartland of E. Lucy Braun's mixed mesophytic forest and found the deciduous forest to have been displaced by boreal forest during the late Pleistocene. No longer tenable was her interpretation of an intact set of tightly coevolved plant communities persisting through millions of years and untouched by climate change. No longer could plant geographers view the many species of the mixed mesophytic forest as having lived together continuously in a central location in eastern North America since the Tertiary Period. Nor could plant ecologists consider that all the other forest types found today across the landscapes of eastern North America were derived from an ancestral mixed mesophytic forest residing on the Cumberland and Allegheny plateaus. It was becoming evident that those forests had assembled in their present composition and geographic distribution sometime in the Holocene interglacial.

Our studies at Nonconnah Creek and Anderson Pond foreshad-

owed many studies that were to follow. But even with these preliminary sites in hand, we knew that evidence was lacking for persistence
of mixed mesophytic forest across what is today the heart of the eastern deciduous forest. With the finds at Nonconnah Creek, we had confirmed that the Blufflands were one major refuge, but the overall area
of the Blufflands is quite small compared to the modern extent of the
eastern deciduous forest. Would there have been enough habitat in just
the Blufflands to prevent extinctions of many of the species we now
consider characteristic of eastern deciduous forest?

Watts's research had showed that southern Florida was no longer
in consideration as a likely refuge for deciduous forest species because
of arid climate and shifting sand dune habitats during the late Pleistocene. His research in central Mexico and that of Bryant and Holloway
in southeastern Texas further indicated that no Texas migration corridor was viable for relocating eastern deciduous forest species during or
after the late Pleistocene. Incredible as it seemed, the only area left
open to the possibility of a Pleistocene refuge for deciduous forest was
on the Gulf Coastal Plain between the Tunica Hills of southeastern
Louisiana and the panhandle of north-central Florida.

The discovery of full-glacial jack pine forest in Tennessee made
us aware that our search for Pleistocene deciduous forest needed to be
focused on an ever-narrowing geographic sector of the southeastern
United States. For the remainder of Paul's research, our quest turned to
finding additional late Pleistocene refuges for temperate deciduous forest species on the Gulf Coastal Plain. To complete Paul's dissertation,
we cored a site in southern Alabama that opened one more window into
the elusive past of the eastern deciduous forest.

139

Pocket Refuges

Up until the era of logging in the late 1800s and early 1900s, the primeval vegetation across much of the outer Gulf Coastal Plain was a savanna-like pine forest in which southern longleaf pine was the dominant species. Longleaf pine grew well on the sandy coastal terraces as well as on the coarse sand and gravel of the Citronelle Formation that forms much of the flat to gently rolling topography of the Gulf Coastal Plain.

Longleaf pine has a very different set of ecological requirements than northern pine species. It is susceptible to ice damage and therefore requires a warm-temperate climate. According to Professor Stuart Ware of the College of William and Mary, longleaf pine thrives in the southeastern United States not just because of the abundant rainfall there but also because wildfires are frequent on the southeastern coastal plains. Longleaf pine is a fire-adapted species, and it requires frequent fires for regeneration. Its cones require heating to high temperatures in order to open and release its winged seeds, which are then dispersed by wind and gravity. The seeds germinate well on a residue of charred ash that provides nutrients on the otherwise dry and nutrient-poor sandy or gravely substrate. Seedlings of longleaf pine grow close to the ground for several years, growing a tuft of long needle-like leaves that help protect the terminal bud from damage. After about seven years in this grass-like stage of growth, the pine saplings bolt upward, rising rapidly above the typical level of ground fires, which under natural conditions burn the wiregrass ground cover every one to three years.

According to the fossil pollen records from Lake Louise in southern Georgia and from numerous lakes in northern Florida, longleaf pine has predominated on the outer coastal plains of the southeastern United States for the past five to six thousand years. Little is known about the changing circumstances that resulted in the rise of the late

141

Holocene southern pine forest, but Bill Watts and others have speculated that it was the result of a combination of rising sea level, increased hurricane frequency, and the activities of prehistoric Native
Americans. With the rise in sea level to its modern position about five
thousand years ago, a seasonal warming of sea-surface temperatures
rose above 80° F, the threshold for the formation of frequent and intense hurricanes, particularly those originating in the Gulf of Mexico.
An increase in storminess, particularly of summer thunderstorms, would
have meant an increase in lightning strikes and a higher incidence of
wildfire. The natural wildfire regime that included a ground fire every
few years probably began when the modern climate pattern was established some six thousand years ago. By that time, Native Americans
may also have been an important ecological factor on the Gulf Coastal
Plain. Prehistoric Native Americans may have set ground fires to drive
game and to increase the line of sight through pine forests for hunting
with atlatls (spear throwers), cane blowpipes, or bows and arrows.

On the outer coastal plain, the terrain is generally flat to gently
undulating and the only fire breaks are streams, lakes, and estuaries
along the coast. As a result, once started, a wildfire easily can burn a
large tract of land before being extinguished. Before the historic era of
logging and fire suppression, many species of hardwood trees were
restricted to river bluffs and wet stream bottoms because they were
intolerant of the frequent, lightning-set or human-set ground fires that
were part of the disturbance regimes across the Gulf Coastal Plain.

Bluffs along the Mississippi, Pearl, Tombigbee, Apalachicola,
and Savannah rivers afforded safe sites for fire-sensitive deciduous
trees because the rivers served as fire breaks. Locally, steep slopes
around sinkhole ponds and bluffs on the lee side of major rivers were
the most important habitats for warm-temperate forest composed of a
mixture of deciduous and evergreen broadleaf trees. In presettlement
times, favorable habitats such as in the Tunica Hills of southeastern
Louisiana and the Apalachicola River bluffs of the central panhandle
of Florida supported extensive tracts of forest dominated by southern
evergreen magnolia and American beech. Today, such sites tend to be
forested with many more species in secondary vegetation that has regenerated on previously cleared land.

In the 1960s, plant ecologists Elsie Quarterman from Vanderbilt University and Katherine Keever of Millersville State College called this secondary vegetation the southern mixed hardwood forest. Quarterman and Keever documented nine tree species as potential canopy formers, with any one stand of trees dominated by several species of coastal plain oak, American beech, southern evergreen magnolia, sweetgum, or black gum (*Nyssa sylvatica*). Originally, stands of beech-magnolia forest and of other southern mixed hardwoods were very restricted in their extent because of frequent wildfires that promoted the spread of longleaf pine. Today, however, southern mixed hardwood forest is becoming more widespread as successional stands of hardwoods are regenerating across the outer coastal plain in areas where natural fires have been suppressed.

When we were students, little was known of the composition of vegetation on the coastal plain before the time of European settlement. No full-glacial sites had been studied, and only the fossil record from the Tunica Hills gave an inkling of the changes in vegetation and climate in the late Quaternary.

Paul reasoned that deciduous forest could have been prevalent on the rolling uplands of the coastal plain, if the climate had been cooler and the incidence of fire had been lower than has been characteristic of the late Holocene. An alternative was the possibility that deciduous forest species were very localized in distribution and confined to small, mesic sites located in ravines and along river bluffs. To test this hypothesis, we first needed to find a fossil pollen site located on an upland site between major river systems. Paul and I therefore made an extensive inventory of possible coring sites across southern Alabama.

Paul chose a shallow pond we called Goshen Springs, located on an old Pleistocene stream terrace well above the modern level of the Conecuh River in south-central Alabama (Figure 25). From examining the topographic map and looking at the site's geologic context, Paul conjectured that the pond probably formed when an old meander of the Conecuh River was cut off many tens of thousands of years ago. The site is now isolated on an extensive area of uplands and surrounded by longleaf pine forest and farmland.

When we arrived at Goshen Springs, the large pond that had

143

Figure 25. Goshen Springs, Pike County, Alabama. (Photograph by Paul A. Delcourt)

showed so prominently on the topographic map no longer existed as open water. The pond had been mostly drained, with a few open pools filled with water lilies dotting the surface of the peatland, which was still mostly wet and not obviously dredged. A big ranch house had been built at the edge of the wetland. With the landowners' permission, we located a large, open pool in the center of the peatland and used our square-rod piston-coring device to core through several meters of reddish-brown, silty and sandy peat, beneath which was ancient stream sand. Because no previous paleoecological studies had ever been made in the vicinity of Goshen Springs, we could only speculate about the age of the sediments or what record of vegetation change they contained.

Most of the peat from Goshen Springs was older than could be measured by standard radiocarbon dating, which typically extends back to about forty thousand years before present. The terrace upon which the pond developed probably dates from the last, or Sangamonian, interglacial interval. The oldest peat in Goshen Springs probably began to accumulate in water-saturated springheads sometime between sixty thousand and one hundred twenty-five thousand years ago. The wetland has been accumulating peat at a very slow rate. Peat accumula-

144

tion rates were so low during the late Wisconsinan glacial maximum, however, that we were unable to get radiocarbon dates directly on the eighteen thousand-year time-line.

Through the uppermost peat, which represented the past twenty to thirty thousand years, changes in pollen percentages occurred smoothly, indicating to Paul that although peat growth slowed during the late Pleistocene, there may not have been a large gap, or hiatus, in the pollen record (Figure 26). Southern pine pollen dominated the pollen spectra from peat that pre-dated the Wisconsinan full-glacial interval. Southern pine was largely replaced by oak through the full-glacial and late-glacial intervals, then once again dominated the pollen assemblage after about eight thousand radiocarbon years before present.

During the last full-glacial interval, the vegetation of the rolling outer coastal plain of southern Alabama apparently was a forest composed mainly of oak, but intermixed with some southern pine. Many additional species of warm-temperate trees were represented through the late Pleistocene in the sediments of Goshen Springs. These trees included hickory, sweetgum, American beech, tupelo gum, and magnolia. Mesic trees such as American beech and southern evergreen magnolia probably grew along the valleys of streams such as the Conecuh River. The predominant vegetation cover over much of the landscape of high river terraces and the Citronelle uplands would have been a more xeric oak-hickory-southern pine forest. Although Paul interpreted his data as evidence that much of the Gulf Coastal Plain of Alabama was dry during the late Pleistocene, he also suggested that, because of the lowered representation of southern pine, wildfires were not as frequent during full-glacial times as in the late Holocene. Consequently, southern mixed hardwood forests were more widespread.

During the late Pleistocene, average annual temperature was probably about the same as today over much of the Gulf Coastal Plain. Although the water level of the Gulf of Mexico was considerably lower during the late Pleistocene than it is today, studies of foraminifera from marine sediment cores by Charlotte Brunner show that the temperature of the sea surface was generally no more than 2° Celsius cooler than in the Holocene — not much change, but just enough cooler to inhibit the formation of hurricanes. The one sector of the Gulf of Mexico that

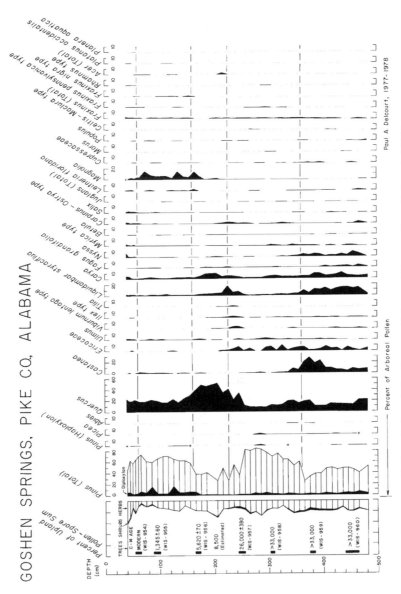

Figure 26. Pollen diagram for Goshen Springs, Pike County, Alabama. (Diagram from Delcourt, 1980)

might have experienced significant cooling was near the mouth of the Mississippi River in Louisiana, where glacial meltwater poured directly into the northern Gulf. In contrast, reconstructions of sea-surface temperatures based upon the work of Bill Balsam and John Imbrie show that the Gulf Stream was displaced to a much lower latitude across the Atlantic Ocean for full-glacial times than during the Holocene. As today, the full-glacial Gulf Stream skirted the western Atlantic coast from Florida north past southeastern Georgia. Instead of reaching as far north as Cape Hateras and sweeping northeastward to Ireland, however, during Wisconsinan times the Gulf Stream flowed directly east from the Carolinas to Portugal. North of about 35° N latitude, the Atlantic Ocean was capped by sea ice.

On land, the ecological transition zone between full-glacial boreal forest and temperate forests must have lain between Bob Black Pond in northwestern Georgia and Goshen Springs in southern Alabama. This interpretation was reinforced by studies of Watts, who, along with Eric Grimm of the Illinois State Museum and Barbara Spross Hansen of the LRC, documented the late-Quaternary fossil pollen records from two sites in northern Florida that contained evidence for full-glacial deciduous forest. At Sheelar Lake, a steep-sided sinkhole pond south of Gainesville, Watts found pollen of southern pine together with that of oak, American beech, and hickory dating between eighteen and fourteen thousand radiocarbon years ago. Camel Lake, a similar sinkhole lake in the Marianna Lowlands west of Tallahassee and close to the Apalachicola River, contained even stronger fossil evidence for mixed deciduous forest during the last full-glacial interval, with large percentages of American beech and oak represented in samples studied by Watts, Grimm, and Hansen. The pollen record from Camel Lake confirmed that limestone sinkholes and phosphate-rich bluffs near the Apalachicola River were Pleistocene refuges for eastern deciduous forest species. In addition, several species of rare and globally endangered endemic plants, such as Torreya (*Torreya taxifolia*), a small cedar-like tree, found only on the Apalachicola River bluffs, probably survived the Pleistocene in these specialized habitats.

Taken together, the fossil records from the Tunica Hills, Goshen Springs, and the Florida sinkhole lakes have helped us to piece together a picture of the landscapes of the late Pleistocene across the Gulf Coastal Plain. Over most of the uplands that are blanketed by sand and gravel deposits of the Citronelle Formation, the vegetation was oak-hickory forest with warm-temperate hardwoods and some southern pine. Warm-temperate swamp species were probably located farther toward the mouths of the major river systems, on exposed stretches of the continental shelf that were inundated during the late Holocene by rising sea level.

Rich, mesic soils in ravines eroded into river bluffs and steep slopes around sinkhole ponds provided pocket-like refuges for many cool-temperate deciduous forest species. These refuges were patchy but generally predictable in their distributions. Each of the major south-ward flowing streams of the southeastern United States may have had locally ameliorated, relatively cool and moist microclimates, and could have served as vital fire breaks. Bluffs lining the Pearl, Tombigbee, Conecuh, Apalachicola, and Savannah rivers, in addition to those of the Blufflands immediately east of the Mississippi River, all could have offered safe havens that sheltered deciduous forest communities through the last glacial-interglacial cycle (Figure 27). Each pocket refuge may have harbored a different assemblage of plants. The biological signifi-cance of this ice-age "archipelago" of island refuges is that different species had different ultimate refuge areas, and these were widely dis-persed across the southeastern coastal plain from the Louisiana Blufflands to southeastern Georgia. Most of the pocket refuges remain elusive because plant-fossil sites in these critical locations are scarce and difficult to locate.

The Blufflands hypothesis, once expanded to form a pocket ref-uge interpretation, represents an intermediate position between the two extreme hypotheses presented by Braun and Deevey while acknowl-edging the overwhelming evidence for world-wide changes in climate during the Quaternary Period. It is now clear that different tree species had different locations for their ice-age refuges and that their founding populations were quite small during much of the Pleistocene. Decidu-ous forest communities that probably had covered much of eastern North America some 125,000 years ago during the Sangamonian interglacial must have disassembled under the stress of climate cooling during the Wisconsinan glacial.

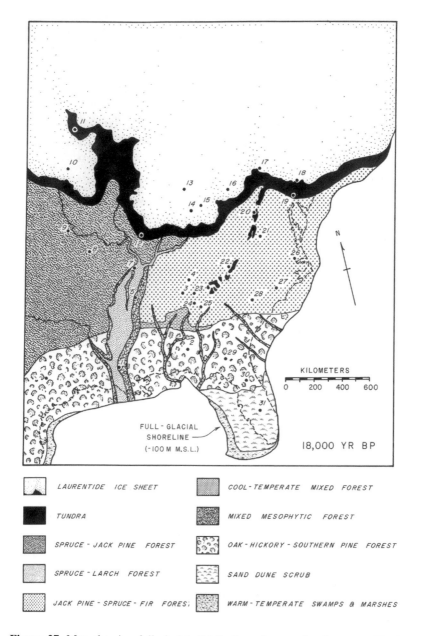

Figure 27. Map showing full-glacial distribution of vegetation in eastern North America, based on sites studied through 1979. Pocket refuges along major streams in the southeastern United States are shown as strips of mixed mesophytic forest pattern. (Modified from Delcourt and Delcourt, 1979)

149

It is the dynamics of the reassembly of biological communities during postglacial climate warming that has created the myriad modern landscape patterns across eastern North America. Reconstructing the full-glacial vegetation patterns (Figure 27) has led to understanding just how restricted the late Pleistocene populations of temperate deciduous trees were. Only during the Holocene did the eastern deciduous forest emerge as a widespread forest formation. By looking at information from all of the pollen sites sampled south of the Wisconsinan glacial margin and recognizing the changes that have occurred in the vegetation of the southeastern United States during the past eighteen thousand years, Paul and I came to appreciate the dynamics involved in the Holocene assembly of plant communities within this major biotic region.

Part IV

Understanding the Present: Predicting the Future

Postglacial Emergence of the Eastern Deciduous Forest

The transition from a glacial interval to an interglacial one is by definition a time of instability during which climate changes rapidly and biological communities undergo major adjustments. Some species become extinct, others adapt, and still more migrate and modify their distributions in response to changing environmental conditions. Our mentor, Herb Wright, suggested that during each glacial-interglacial transition, it is as if the "ecological cards are shuffled." By this he meant that in a given interglacial, biological communities arise that are different from those of previous glacial and interglacial times in both the sequence in which they assemble and in their composition. One reason for this is the element of chance, the particular circumstances that determine the locations of glacial-age refuges from which species spread as the interglacial assembly gets underway. Another factor is the migration abilities of different species, which vary according to the type of seeds produced and the ability of wind, water, or animals to disperse them.

Each interglacial interval is a unique combination of climate conditions resulting from the complex interplay of astronomical linkages between changes in the earth's orbit about the sun, the so-called "Milankovitch cycles." From an anthropocentric perspective, we tend to think of the Holocene, our interglacial, as the norm, but it is actually only the most recent in a sequence of many interglacials that occur at one-hundred-thousand-year intervals. The Holocene is unique because we now lack many species of large mammals, or "megafauna," that became extinct at the end of the Pleistocene. In their absence, new mammal communities have developed, vegetation has reorganized, and human civilization has arisen. Human cultural interactions with Ho-locene biotic communities have intensified worldwide during the Holocene, resulting in increasing amounts of the land surface being converted from natural to cultural landscapes.

After graduating from the University of Minnesota in 1978 and setting up shop at the University of Tennessee, we, with our graduate students, continued to expand our information base across the southeastern United States through a series of studies sponsored by the National Science Foundation. One goal was to produce maps of changing vegetation on a grand, sub-continental scale. We dubbed this project "FORMAP — Forest Mapping across Eastern North America," with the aim of reconstructing the changes in forest vegetation at individual sites across the region, and then mapping those changes, both for individual species and for plant communities from full-glacial times to the present. As part of the FORMAP project, published in our 1987 book, *Long-Term Forest Dynamics of the Temperate Zone*, we explored the ecological implications of our findings, emphasizing pattern and process in the emergence of regional forest communities through the Holocene interglacial.

From our studies and those of our colleagues across the temperate zone, it was apparent that certain Pleistocene biotic communities, for instance the "mammoth steppe" ecosystem of eastern Siberia and western Canada, ceased to exist after the transition to interglacial conditions. Vegetation units we recognize as having been widespread before the most recent disruption with European American settlement, including the prairie grasslands of the Great Plains region and the eastern deciduous forest, occupied only very small areas or were disaggregated with individual species intermingling within other communities during the Pleistocene. Vegetation types we recognize today emerged in their present form only in the past ten thousand years. By "emergence" I mean not only the expansion of species ranges to their present distribution but also the assembly of communities of plants and animals to their present composition. The eastern deciduous forest emerged as a major plant formation only after climate began to warm some sixteen thousand years ago. This emergence of deciduous forest was the result of the outward spread of constituent tree species from small, isolated pocket refuges. The processes responsible for the changes in location, area, and composition of deciduous forest communities included changes in climate and soils that took place over thousands of years as species gradually adjusted to Holocene interglacial conditions. What we sought to study was not only the timing of adjustments in ranges of individual species and the rates of change in their distributions, but the patterns of change in community composition as new combinations emerged from this most recent "reshuffling" of the ecological card deck.

In our research, we wished to find paleoecological sites that would fill gaps in previous knowledge. These new sites would provide vital links, helping us to pinpoint the Pleistocene locations of tree species and to trace changes in their whereabouts during postglacial times. Also crucial to the FORMAP project was developing numerical methods to translate fossil pollen percentages into estimates of past abundance in the vegetation. From these estimates, rates of change could be calculated for both species ranges and community composition, and this formed the basis for a series of maps depicting the general patterns of vegetation change after the Pleistocene.

One region that we needed to investigate was the several-hundred-kilometer gap between middle Tennessee and the late Wisconsinan glacial margin near the Ohio River. Did jack pine forest extend from the coastal plain of the Carolinas across the Piedmont of northern Georgia, west to the Mississippi River, and north to the southern margin of the Laurentide Ice Sheet? Or did pine forest give way to spruce forest in the northernmost full-glacial boreal forest, as it does today in central Canada? Was there treeless tundra vegetation south of the glacial margin in Kentucky or did forest grow right up to the edge of the glacial ice margin? To answer these questions, we needed to find and study a representative site in north-central Kentucky.

The Mississippi River Valley was a second area in need of more study, in part to give us a better understanding of the gradients in the composition of full-glacial boreal forest, but also to further test our hypothesis that the Blufflands had been a restricted refuge for deciduous forest species. What was the vegetation like on the wind-swept, foggy point-bar environments of the vast braided stream that was the full-glacial Mississippi River? Was temperate hardwood forest to be found there on the channel flats, or were they occupied by a vast boreal woodland or a treeless plain? A strategy was needed to find a full-glacial pollen site within the braided-stream channels of southeastern Missouri or northeastern Arkansas.

The Ozark highlands were also largely uncharted but critical to understanding the southern bounds of the late Pleistocene boreal forest, as well as the sequence of immigration of deciduous forest species northward following climatic amelioration in the late-glacial interval. Was spruce more important west of the Mississippi River than to its east? Were the Ozark highlands a possible refuge for temperate deciduous forest species? Sinkhole ponds on the limestone plateaus of southeastern Missouri held the promise of answering these questions.

Finally, closer to our home base in Knoxville, many questions remained about the significance of the southern Appalachian Mountains in the biogeographic history of the southeastern United States. The southern Appalachians, including the Great Smoky Mountains of eastern Tennessee and western North Carolina, today harbor a rich mix of native species of vascular and nonvascular plants, vertebrate and invertebrate animals, and fungi, including not only boreal and temperate forms but also species with subtropical and tropical affinities. Many species are endemic to the southern Blue Ridge Province and have limited dispersal abilities — where did they survive during the Pleistocene?

The FORMAP project and its related spin-offs became a major four-dimensional puzzle across space and through time. As we began to locate key paleoecological sites, however, many of the pieces began to fit together into a coherent picture of past landscapes. The remainder of this chapter describes those key findings and concludes with a map summary of changes in the vegetation regions of eastern North America since the last glaciation.

Some twenty thousand years ago, the great Laurentide Ice Sheet stopped its advance just south of the present location of Dayton, Ohio, and built glacial moraines there while leaving northern Kentucky free of ice (Figure 27). Not one Pleistocene pollen site had been studied in the state of Kentucky, however, until 1983 when Gary Wilkins, one of our graduate students, began to search for ancient lakes in the Bluegrass region of north-central Kentucky. On a farmstead located about forty kilometers north of Mammoth Cave National Park, Gary found a prospective site he called Jackson Pond. The site is located at the juncture of two physiographic provinces, where a prominent line of low hills marks the Dripping Springs Escarpment, the boundary between the Cumberland Plateau to the east and the Western Coal Field region to the west.

Using a topographic quadrangle printed in the 1950s, Gary located several sinkhole ponds northwest of the Dripping Springs Escarpment near Mammoth Cave National Park. Upon interviewing the owner of the land upon which these ponds were located, Gary discovered that between 1950 and 1983 almost all of the ponds had been disturbed. Mr. Jackson had used sticks of dynamite to deepen all of his ponds for cattle watering holes — only one pond on his farm remained with its sediments intact, and that one, which Gary called Jackson Pond, was kept as a natural fishing hole. Mr. Jackson gave Gary permission to collect and study cores of sediment from Jackson Pond.

Jackson Pond is about the size of a football field. A few trees, mostly sweetgum and swamp willow, ring its perimeter, along with some buttonbush shrubs. Shortly after our coring crew, composed of Gary, Paul, me, Dave Shafer, and Newman Smith, assembled to core Jackson Pond, I ventured out into the water at the pond's edge. The overall visual impression of the pond is of a solid green surface, because the surface is covered by a continuous floating leaf mat of yellow water lily (*Nuphar tuberosa*). These yellow-flowered aquatic plants grow in water that is no more than a meter deep. They must remain rooted in soft mud and yet send up their stems to the water surface to catch the sunlight. With no emergent plants creating above-water barriers, it appeared at first that it would be easy to float our raft to the center of the pond, but the lilies soon proved to be a formidable, spongy barrier to our progress. Once the neoprene rafts were pumped up and the coring platform was in place, we pushed our raft slowly out over the surface of the water lily mat. In the center of the pond, one small opening in the mat of water lilies, about three times the diameter of the coring rig, offered a place to float free of the rope-like water lily stems so that we could operate the square-rod piston sampler free of tangles. Based on his earlier probes with a small diameter soil sampler, Gary expected Jackson Pond to yield about six to seven meters of organic-rich clay, similar in accumulation of sediment to Anderson Pond, located some two-hundred-fifty kilometers to the south.

The Jackson Pond sediment record proved comparable to that

from Anderson Pond in more than its total thickness. The deepest sediment samples from Jackson Pond were dated at twenty thousand radiocarbon years ago. The fossil pollen assemblage from full-glacial sediments was composed mostly of jack pine pollen, but with a higher percentage of spruce pollen than at Anderson Pond. No evidence of tundra herbs was found at Jackson Pond, indicating that a mixed boreal forest of spruce and jack pine grew close to the ice sheet when it stood at its maximum. Tundra, which today occupies a broad latitudinal zone across the Arctic Circle, must have been confined to a very narrow belt just along the ice front during the late Pleistocene. The changeover from boreal forest to deciduous forest occurred much later at Jackson Pond than it did at Anderson Pond, however. Oak and other deciduous trees did not become important at Jackson Pond until ten thousand years ago, some six thousand years later than at Anderson Pond. Evidently, climate warmed first farther south and migrations of deciduous forest species were delayed to the north because of unsuitably cold climate conditions that still prevailed in the Midwest.

West of Jackson Pond, and south of the confluence of the Mississippi and Ohio rivers, lies the vast Lower Mississippi River Valley. Because of its complex history during the Quaternary, the Lower Mississippi Valley consists not only of modern meander trains but also of older surfaces that were once active river channels fed by melting glaciers that are now abandoned terrace surfaces. During the Wisconsinan full-glacial interval, the braided Mississippi River traversed a path across the western part of its valley, known as the Western Lowlands. These old braided-stream terraces lie west of a low ridge of sand and gravel deposits of Tertiary age called Crowleys Ridge. Some of the low swales carved out between the point bars of the braided streams became small lakes during the late Pleistocene. A very few of those small lakes have collected sediment over as much as the past twenty thousand years. One such place is Powers Fort Swale, a site called to our attention by Jim and Cynthia Price.

Powers Fort Swale is located in the Western Lowlands of southeastern Missouri. Its name derives from its proximity to a large fortified Indian village dating from about AD 1350. Today Powers Fort Swale is a small, quiet-water pond ringed with large bald cypress trees. Dan Royall studied the pollen record from Powers Fort Swale with us as part of his Master's thesis. What Dan Royall discovered at Powers Fort Swale was a late Pleistocene environment quite unlike that to the east in the Blufflands or at Anderson and Jackson ponds. He found that the vegetation growing on the shifting sand and gravel bars of the braided stream channels consisted mainly of spruce and willow, with some tamarack but no jack pine. Willow shrubs would have been well adapted to withstand frequent floods in the meltwater channels of the braided river, because they can establish themselves on gravelly point bars and can resprout from roots. The spruce, the most abundant tree on the Western Lowlands during the late Pleistocene, was probably the now-extinct *Picea critchfieldii*, which apparently was well adapted to the cool, foggy floodplain environments of the Pleistocene Western Lowlands.

With assistance from David Anderson of the Southeast Archaeological Center, Tallahassee, Paul and I studied Hood Lake, a site located near Jonesboro, Arkansas. At Hood Lake, the modern swamp forest composed of bald cypress, tupelo gum, and other warm-temperate wetland trees did not become established until after cold glacial meltwater ceased to flow from the southern margin of the Laurentide Ice Sheet. The changeover in the Mississippi River from a braided stream to a meandering one occurred about ten thousand radiocarbon years ago, and it marked the onset of the Holocene interglacial in the Western Lowlands.

West of the Mississippi River Valley, Newman Smith's studies of Cupola Pond sediment (figures 1, 3) confirmed the observations of Mehringer and King at Boney Spring on the Pomme de Terre River of western Missouri. The oldest sediment recovered from Cupola Pond was deposited seventeen thousand radiocarbon years ago and contained a pollen assemblage dominated by spruce with negligible amounts of jack pine. As at Boney Spring, spruce was the most important tree in full-glacial forests on the watershed of Cupola Pond. And as at Boney Spring, pollen grains of deciduous forest trees were scarce during the Wisconsinan glacial maximum. The deciduous forest must have been displaced south of the Ozark highlands of Missouri during the late Pleistocene.

Cupola Pond is significant for its record of the changes in vegetation that transpired during late-glacial and postglacial time. Newman Smith's study showed that the changeover from late Pleistocene spruce forest to Holocene oak-hickory forest in the Ozark highlands involved a complex and long series of local extinctions, immigrations, and changes in the relative importance of tree species on the landscape. This study lends insight into the individualistic nature of species responses that result in dynamic changes in forest composition during times of major climate change.

From the combined studies of Frey, Whitehead, and Watts in the Carolinas and Georgia, of Mehringer and King in the western Ozark highlands, and of our group's studies in the intervening territory of Tennessee, Kentucky, and Missouri, an overall picture became clear for full-glacial vegetation of the region immediately south of the continental ice sheet in eastern North America. In the late Pleistocene, boreal-like forest covered an extensive area (Figure 27). In Tennessee, Georgia, and the Carolinas, the boreal-like forest was dominated by jack pine. To the north and west, spruce became increasingly important. Only a few cold-hardy deciduous trees were recorded by pollen grains. These included black ash, which grows today north into the southern boreal forest, and willow, which may have been a shrub willow not unlike those species that today grow above the Arctic Circle. Eastern jack pine forests may have been quite open and species-poor, particularly on the sandy coastal plain of the Carolinas. Even in Tennessee, there is evidence that the jack pine forest did not develop a dense closed canopy. Minerals that compose the slopes of the Cumberland Plateau over a kilometer away from Anderson Pond were responsible for the high sediment accumulation rates at Anderson Pond during full-glacial times. Reconstructions of open forest interspersed with open-ground vegetation are consistent with fossil evidence of prairie pocket gophers (*Geomys bursarius*) found by professors Walter Klippel of the University of Tennessee and Holmes Semken of the University of Iowa from faunal sites in Arkansas, Missouri, and eastern Tennessee that date from the Wisconsinan glacial maximum.

In the central and southern Appalachian Mountains, Watts demonstrated through pollen and plant macrofossil analysis of sediments from a series of bogs and ponds that the full-glacial vegetation of high elevations was alpine tundra dominated by sedges, dwarf shrubs, and arctic-alpine herbs. That this Pleistocene alpine tundra occurred on the

Allegheny Plateau as far south as Maryland and West Virginia was demonstrated by Maxwell and Davis at Buckles Bog, Watts at Cranberry Glades, and our student Peter Larabee at a site called Big Run Bog, a peat bog perched on nearly flat-lying sandstone. Cored with the help of Professor Gerald Lang of West Virginia University, the two meters of peat underlying the bog surface were found to have accumulated over the past seventeen thousand years. Pete's studies of changes in peat and clay sediments, along with the fossil pollen record, showed that changes in climate at the end of the Pleistocene triggered changes in the entire landscape, with changes in the soil as well as the species of plants growing around Big Run Bog. Full-glacial climates in the central Appalachian Mountains were severe, and frozen ground thawed only briefly during the summer. As the climate warmed, seasonal freezing and thawing became intense and the land surface became unstable. Sediments that had been locked up on the perennially frozen mountainsides began to move downhill in gullies and streambeds. Plant cover was sparse during the late Pleistocene. Mosses and other small clumps of "cushion" plants dotted the landscape. These tundra plants survive the severe cold of arctic regions today. When the climate warmed, spruce and balsam fir trees moved in, first forming sparse, stunted shrubs and then growing to full tree height as conditions became more favorable. The changeover from alpine tundra to forest took place at the same time as the perennially frozen ground thawed and soil began to develop on the watershed of Big Run Bog, between fourteen and ten thousand years ago. Big Run Bog still has red spruce growing around its margin, although the uplands beyond the bog are today clothed in a mixed deciduous forest of American beech, sugar maple, black cherry (*Prunus serotina*), and other hardwoods.

Farther south, on the summits of the ten highest peaks of the southern Appalachian Mountains, direct evidence of full-glacial vegetation is lacking. There, the topography is more rugged, and intermountain valleys suitable for bog development are scarce. Wetlands such as those in Long Hope Valley, located in the northwestern corner of North Carolina, harbor many species of northern plants, such as the northern yew (*Taxus canadensis*) and bog buckbean (*Menyanthes trifoliata*) that today reach their southern limits in the southern Appalachian Mountains. We probed the Long Hope Valley bogs with the assistance of Professor J. Dan Pittillo of Western Carolina University in the hope of discovering whether these plants are true relics of the Pleistocene that have persisted in these high-elevation wetlands because of cold-air drainage which even today modifies the otherwise warm climate. We found that wetlands such as those in Long Hope Valley generally contain organic-rich sediment records extending back no more than a few thousand years because of continual reworking by streams that meander across the peatlands. Nevertheless, the wetland habitat must have persisted since the late Pleistocene to account for the present distributions of so many northern plant species.

Other, indirect evidence of Pleistocene tundra landscapes was found at Flat Laurel Gap, North Carolina, by our student Dave Shafer. Flat Laurel Gap is a bowl-shaped basin located along the Blue Ridge Parkway near the summit of Mount Pisgah. Dan Pittillo had suggested Flat Laurel Gap as a study site both because of its shape, which was similar to cirque basins in the Rocky Mountains that were carved out by glacial ice during the late Pleistocene, and because of the extensive heath bog and sedge wetland within the basin. Dave discovered that Flat Laurel Gap contained a plant-fossil record extending back for only the past three thousand years. Surrounding the bog, however, he found abundant geologic evidence, in the form of now-immobilized boulder fields and colluvium (a mixture of cobbles, sand, and silt), that perennially frozen ground with permanent snow packs dominated this landscape some twenty thousand years ago. Boulder fields form today under conditions of intense freeze-thaw activity that literally plucks the rocks from the hillside and concentrates them in linear piles extending from hilltops to valley bottoms. Colluvial deposits formed during the transition from glacial to interglacial conditions, as the soil warmed and a water-saturated slurry of sediment flowed downslope. Similar deposits are found at elevations above fifteen hundred meters throughout the southern Appalachians, leading to the interpretation that at least the ten highest peaks were above treeline during the late Pleistocene.

Our search for Ice-Age Appalachian springs took us not only to high-elevation bogs but also to wetlands located in valleys at lower elevations. Paul and I found park service official Gary Larsen eager to

162

give us permission to look for permanent springs, ponds, and bogs in the Great Smoky Mountains National Park, but we found little there that was suitable for our work because the mountainsides are so rugged and well-drained. We were intrigued, however, by one possibility called Lake in the Woods, and with ecology graduate student Jean Davidson we set out to explore its history. Lake in the Woods is located in Cades Cove, a geologic window or large valley underlain by relatively soft limestone in an area otherwise composed of metamorphosed sedimentary rocks that are much more resistant to weathering. Several natural ponds existed in Cades Cove at the time of pioneer settlement in the early 1800s. Only Lake in the Woods has survived in its original condition, with buttonbush shrubs fringing the shallow pond and a canopy of sweetgum trees arching overhead.

A Cherokee legend tells of a "vanishing lake" in the middle of Cades Cove. This story described Lake in the Woods well, for it expands to several hectares with the winter rains and shrinks to a small puddle in the heat of summer. Jean's pollen record showed that the pond had not been wet throughout the year at any time in the recent past. It seems to have expanded to its largest extent about six thousand five hundred years ago when climates of this part of the southern Appalachians were warm and rainfall was even more abundant than it is today. Lake in the Woods became a wetland in the mid-Holocene interval, which can explain why rare warm-temperate species grow there today but does not help explain what Ice-Age conditions were like.

For the key to understanding late Pleistocene climates at low elevations in the southern Appalachian Mountains, we turned to studying a fifteen-thousand-year sediment sequence from Saltville Valley in southwestern Virginia. Saltville has been famous for its Pleistocene mammal remains since the 1700s, when a mastodon tooth discovered there was sent to Thomas Jefferson, who was an avid fossil bone collector. Over the past century, the gray clay deposits of the ancient Saltville lake bed have yielded fossil caribou (*Rangifer tarandus*), woodland musk oxen (*Bootherium bombifrons*), and other species that today are either extinct or else are confined to high latitudes in boreal and arctic climates. In 1982, Paul and I accepted an invitation from Professor Jerry McDonald of Radford University, to visit his excavations at this famous site. Our task was to collect a set of samples for pollen analysis from a section of gray clay exposed in a trench cut with a backhoe. The gray clay deposits represented a former lake bed and overlay stream gravels containing fossil bones of a woodland musk ox. The fossil bones were preserved where the animal had died in the ancient lake, its carcass buried by lake clays that settled over it. The ancient lake had formed as the valley outlet was dammed, probably by a late Pleistocene landslide or flowing blockstream. Our samples were clay blocks cut with a trowel from the two meters of clay bank cleaned away above the fossil bone bed.

McDonald's musk ox and Jefferson's mastodon lived in an Ice Age landscape that was less rigorous than that of the high mountain peaks of the Appalachians. At Saltville, fifteen thousand years ago, the vegetation was a mixture of cold-tolerant spruce and more temperate oaks and other hardwoods. Locally the margins of gravel-bottomed streams were grassy, with shrubs of green alder (*Alnus crispa*) growing in streamside thickets. Based on the types of plants in the fossil record, the glacial-age landscape was a mosaic of forest and open marshes bordering Appalachian springs that would have provided abundant browse for many animals. Later, by twelve thousand years ago, eastern hemlock trees formed dense stands on the mountainsides surrounding Saltville Valley. Modern forests of chestnut and oak replaced eastern hemlock after nine thousand years ago. Today, the boreal green alder shrub persists in this part of the Appalachians only on the summit of Roan Mountain, straddling the Tennessee/North Carolina state line. That one relict population of boreal green alder serves as a reminder of the long-lost Ice Age landscapes of the southern Appalachian Mountains.

Twenty thousand years ago, the summit of Mount LeConte and nine other high peaks in the southern Appalachians were blanketed by alpine tundra plants interspersed with permanent snowfields (Figure 28). Below fifteen hundred meters elevation, the lower limit of red spruce and Fraser fir today, lived boreal-like conifer forest, which spread down the mountainsides and outward across the adjacent Ridge and Valley physiographic province to the northwest in Tennessee and Virginia and the Piedmont to the east in the Carolinas. After sixteen and a half thousand years ago, when climates began to warm, patches of tundra and

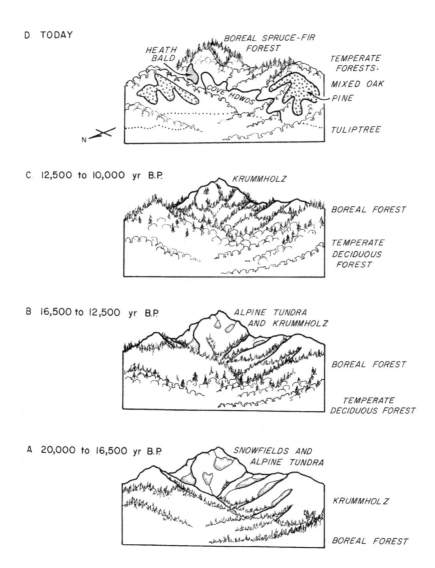

Figure 28. Changes in vegetation on Mount LeConte, Great Smoky Mountains National Park, over the past twenty thousand years. (Diagram from Delcourt and Delcourt, 1987)

snowfields shrank and conifers ascended to higher elevations. By twelve to ten thousand years ago, periglacial landforms stabilized, streams began to flow in well-defined channels down the mountains, and tundra plants would have been restricted to bare rock outcrops, landslide scars, and intermontane wetlands. Deciduous forest began to spread upslope and to migrate to more northern latitudes. Only after ten to eight thousand years ago did the vegetation patterns we recognize today become defined.

The dynamic patterns and processes we envisioned for the southern Appalachian Mountains, along with the many complementary site studies undertaken across the southeastern United States during the 1970s and 1980s, helped us to prepare maps depicting the broad patterns of change in vegetation across both elevational and latitudinal gradients (Figure 29).

The general pattern of emergence of Holocene deciduous forest can be visualized from our sequence of maps from eighteen thousand years ago to the present (Figure 29). During the Wisconsinan glacial maximum, most of the area of eastern North America now characterized as eastern deciduous forest was covered by glacial ice, treeless tundra, or boreal forest. A sharp transition zone at about 34° N latitude separated boreal from temperate vegetation. The Blufflands and other pocket refuges scattered across the southeastern coastal plains supported local stands of mixed mesophytic forest. Elsewhere across the Gulf Coastal Plain forests were a mixture of dry oak-hickory and southern pine forest. The Florida Peninsula was largely covered by dry, shifting sand dunes and scrubby vegetation. Warm-temperate swamp forest may have existed on the continental shelf along river valleys that were submersed by the Gulf of Mexico as sea levels rose during the Holocene.

Increases in area of deciduous forest began after sixteen thousand years ago following the first melting back of the southern margin of the Laurentide Ice Sheet and the dieback of spruce and jack pine forests in middle latitudes. Tundra and boreal forest tracked the retreat of the ice northward, and deciduous forest communities began to assemble as black ash and hornbeam expanded in place from their position in the full-glacial boreal-like forest region. Oak and hickory spread generally northward from the Gulf Coastal Plain. American beech, elm, walnut, tulip tree, maples, and other mesic hardwood species each increased from relatively restricted and isolated refugial populations to add to the changing blend of deciduous forest communities during the late-glacial interval of climate warming from sixteen to ten thousand years ago.

By ten thousand years ago, prairie plants had expanded and coalesced into a coherent vegetation region spreading eastward from the Great Plains and forming a discrete boundary with forests at about 95° W longitude. Over much of eastern North America, forest became temperate in character as tundra and boreal forest moved northward into Canada. After eight thousand years ago, prairie reached its maximum eastward extent, then began to recede to the west. Warm-temperate swamp forest filled the Lower Mississippi Valley as well as other stream courses throughout the southeastern United States. Mixed mesophytic forest, oak-hickory forest, and Appalachian oak-chestnut forest assumed their modern positions in the heartland of the eastern deciduous forest. Adjustments in boundaries of vegetation types continued through the late Holocene in response to gradual climate cooling, but the present configuration of the eastern deciduous forest region largely took shape by about five thousand years ago in the middle Holocene.

Changes in percentages of oak trees (*Quercus*) over the past eighteen thousand years exemplify the individualistic nature of species responses to climate change during the transition from glacial to interglacial climates (Figure 29). During the Wisconsinan full-glacial interval, temperate species of oak were restricted to a narrow zone along the Gulf Coastal Plain south of 34° N latitude, where they composed up to twenty percent of the forest vegetation. Within five hundred to fifteen hundred years after the beginning of the retreat of the Laurentide Ice Sheet, oaks began to advance northward, as evidenced at sites such as Bob Black Pond, Georgia, Anderson Pond, Tennessee, and Cupola Pond, Missouri. By fourteen thousand radiocarbon years ago, populations of oak trees increased to forty percent of the forest composition in the southeastern coastal plains and represented twenty percent of the forest from northwestern Georgia, across Tennessee, to the Ozark highlands of Missouri.

By ten thousand radiocarbon years ago, the beginning of the Holocene interglacial, oaks had spread not only northward, but southward

A

B

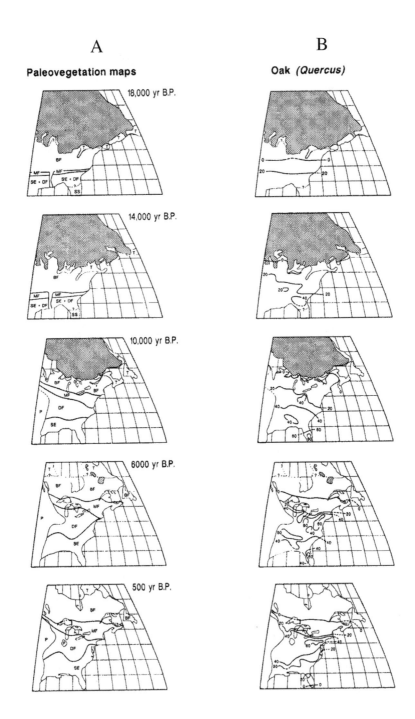

Paleovegetation maps

Oak *(Quercus)*

18,000 yr B.P.

14,000 yr B.P.

10,000 yr B.P.

6000 yr B.P.

500 yr B.P.

168

as well, becoming more than fifty percent of the forest of peninsular Florida and expanding to forty percent of the forest in the southern Appalachian Mountains. The northern limit of distribution of oak trees in the early Holocene reached nearly to the southern margin of the re- treating Laurentide Ice Sheet, which was by then positioned across the Great Lakes region. By six thousand radiocarbon years ago, oak forest dominated across the breadth and width of the modern-day eastern de- ciduous forest region, and several oak species even extended beyond their modern range limits into southern Canada during the mid-Ho- locene time of maximum climate warming. Southern pine forest dis- placed oak forest on the Gulf Coastal Plain, and drought-tolerant oak species spread westward into the prairie region. At the time of first cultural contact by the Spaniards some five hundred years ago, oak was quantitatively the most important group of tree species of the east- ern deciduous forest. It was for this reason that foresters Shantz and Zon characterized the great eastern deciduous forest region as "essen- tially fagaceous in nature" — that is, dominated by species within the plant family Fagaceae, which includes both oak and American beech.

To understand the complexity of the processes by which the east- ern deciduous forest emerged during the present interglacial, we need to look back at the circumstances under which temperate forest species survived during the Wisconsinan. Altogether, mixed mesophytic forest species were restricted to an area less than one-tenth of their present distribution. Different species found refuge in different pockets of habitat that were widely dispersed across the Gulf and southern Atlantic coastal plains. Some of these species remained restricted through the Holocene and are now considered to be narrowly distributed endemics, while others were destined to become widespread species.

Figure 29 (opposite page). A. Glacial-interglacial changes in extent of glacial ice (shaded pattern) and vegetation mapped across eastern North America for the past eighteen thousand years (from Delcourt and Delcourt, 1993). Vegetation types are as follows: T = tundra, P = prairie, SS = sand dune scrub, BF = boreal forest, MF = mixed conifer-northern hardwoods forest, DF = deciduous forest, SE = southeastern evergreen forest. **B.** Maps of changing distribution and relative abundance of oak (*Quercus*) across eastern North America for the past eighteen thousand years. Con- tours represent percent of reconstructed forest composition. (Diagram from Delcourt and Delcourt, 1993)

Twenty thousand years ago, populations of deciduous forest species were small and relatively isolated from one another. Such geographic isolation can affect a species in either of two ways, depending
upon its ability to adapt to new opportunities. Some species are inherently conservative. They exchange genetic material only within a local
population and are not broadly tolerant of environments that change
widely through space or over time. An example of this type of genetic
conservatism is found in the genus *Juglans*. Both black walnut (*Juglans
nigra*) and butternut (*Juglans cinerea*) are tree species considered characteristic of the mixed mesophytic forest today, but they tend to be
restricted to favorable mesic slopes and are not abundant within the
mixed deciduous forests in which they are found. When carried to the
extreme, genetic conservatism leads to specialization that may reduce
the population of a species to such low levels that it becomes vulnerable to extinction. Narrowly endemic species such as the Florida Torreya
and the Franklin tree are cases in point. In the former case, the species
grows today only on a few bluffs along the Apalachicola River where it
is protected from human disturbance. Unfortunately, the only known
naturally occurring grove of wild *Franklinia* trees, located by William
Bartram along the Altamaha River in southeastern Georgia, has disap-
peared; the species exists today only in cultivation in plant nurseries
and gardens. Some narrowly endemic species, such as Kentucky coffee
tree (*Gymnocladius dioicus*) and arrow-wood (*Maclura pomifera*), with
their large edible fruits, now may be restricted in distribution because
they have lost their primary megafaunal vectors, mastodons and other
large herbivores, through extinction at the end of the Pleistocene.

A second kind of genetic makeup, characteristic of many tree
species, allows their populations to have resilience in the face of alter-
nate restriction and isolation during times of unfavorable climate, followed by release from small refuge areas with amelioration of climate.
This characteristic is called "ecophenotypic plasticity." It is the ecological expression within the living population (the phenotype) of a
diverse genetic makeup. Individuals within such a population inter-
breed freely and their offspring are highly adaptable to new situations.
Sugar maple is widespread within the eastern deciduous forest and is a
key species of several forest communities, including not only the mixed

mesophytic forest but also the beech-maple and maple-basswood forests. Sugar maple is a complex of a number of subspecies and exhibits
wide variability due to its genetic plasticity. The forty or so species of
oak are famous for their ability to hybridize with one another, which
makes identifications in the field often difficult but probably accounts
for their success in occupying a central role in the Holocene eastern
deciduous forest.

In addition to genetic diversity within populations, the life-history characteristics of tree species greatly influence their ability to adjust their ranges during times of climate change. Because individual
plants must remain firmly rooted in the ground, a critical factor in
changing the geographic range of a species is its means of dispersal of
propagules such as seeds, fruits, or cones. Different species have different strategies for migration based upon their dispersal abilities. If
the distance between late Pleistocene pocket refuges exceeded the dispersal distance of pollen grains or seeds, interchange of genetic material was poor, resulting in reduction in the diversity of the gene pool for
some species. With climate warming, those species that were generalists, that is, broadly tolerant of climatic fluctuations, were favored over
specialists adapted to specific and narrow environmental conditions.

During the late-glacial and Holocene interglacial intervals, generalist tree species such as aspen and white spruce were the first to
spread northward in the wake of the melting continental ice sheets.
These generalists produce winged or tufted seeds that can disperse fast
and far, blown by the wind across nonforested landscapes of barren
tundra, snow, or ice to sites suitable for seed germination. These trees
are sometimes called "r-selected" or "pioneers," that is, they are short-
lived and produce pollen and seeds at a relatively young age, and they
can grow on sites with bare soil newly exposed by melting glacial ice.
They are also more cold-hardy than most trees, a physiological condition that permits successful colonization of populations whose seeds
disperse long distances into often extreme environments. Behind the
northward advancing front of pioneer species came a second wave of
migrants composed of species with slower growth rates and more localized dispersal capabilities that required mature soils enriched in ni-
trogen and accumulating organic matter as well as more stable and

more temperate climates. These "k-selected" or "late-successional" species include oak, American beech, and sugar maple.

The process of spreading out from limited "founder" populations in pocket refuges to ever-expanding territories is a slow one in the case of trees. As much as twenty-five to fifty years may be required from germination of an individual seed to first pollination and seed set. Dispersal vectors may advance the forest edge only by a few tens of meters a year. Nevertheless, when we calculate the rates of advance of deciduous forest trees northward during postglacial times, we find that the pace of expansion was remarkably swift — up to several hundred meters per year on average. Margaret Davis, Professor Jim Clark of Duke University, and we have discussed the implications of this rapid movement in a series of publications. Such fast rates of tree migration as observed in the Quaternary fossil record cannot be explained by a simple diffusion process where a species spreads out from a central point like concentric ripples on a pond onto which a stone has been tossed. Instead, relatively infrequent but long dispersal events tend to result in outlier populations of species well beyond the general margin of the advancing population front, and these founder populations then reproduce and help to accelerate the overall process of species migrations. Rapid advances of the colonizing fronts for both aspen and spruce have been reported in New England and across central Canada. In these cases, the migration fronts advanced by hundreds of kilometers in a single "jump dispersal." For oaks and American beech, the rates of advance were slower, on the order of several kilometers per year, and most were facilitated by blue jays (*Cyanocitta cristata*) and small mammals. Margaret Davis and Sarah Webb have suggested, however, that American beech crossed certain significant geographic barriers such as Lake Michigan in long-distance dispersal events of between twenty-five and 130 kilometers, probably carried over these barriers by avian vectors such as blue jays or the now-extinct passenger pigeon (*Ectopistes migratorius*).

To the extent that the adaptations, life-history strategies, and reproductive and dispersal capabilities of different tree species are different from one another, the species will tend to be distributed individualistically, and they will respond to climate change individually in different ways. This was the premise initially put forth by Henry Gleason of the New York Botanical Garden in his 1926 article on the individualistic nature of the plant association. Through the work of generations of Quaternary paleoecologists, not only those studying eastern deciduous forests but also those working in such diverse environments as the Grand Canyon of Arizona, the Outback of Australia, and the Amazon Basin, Gleason's premise is now upheld as one of the fundamental paradigms in contemporary plant ecology. During the full-glacial interval in the southeastern United States, life zones of eastern North America were not simply compressed into a much smaller space. Some plant communities remained similar to those we see on today's landscape — excellent modern analogs can be found for Pleistocene boreal jack pine and spruce forest. Other Pleistocene plant communities, however, were unlike any in existence today.

Even more startling to Paul and me, however, was the discovery that during times of transition between glacial and interglacial climates, ephemeral biological communities formed that neither resembled those of full-glacial times nor those of full-interglacial times. This discovery became clear as we mapped the former distribution of two species that today are widespread within the eastern deciduous forest but only as minor constituents. During the late-glacial and early Holocene intervals, however, both species of hornbeam (*Ostrya virginiana* and *Carpinus caroliniana*) underwent a dramatic rise, and then decline, in abundance throughout a geographically significant sector of eastern North America. My interest in their story began with the pollen record from Anderson Pond, where, embedded within a long sequence of pollen samples dating from about thirteen to nine thousand radiocarbon years ago, was an unusual occurrence. I found that pollen percentages of hornbeam increased systematically over time. By twelve thousand radiocarbon years ago, it was thirty percent of the tree pollen. Such a population explosion in hornbeam had not been recorded before from the Pleistocene/Holocene transition.

Was the rise and decline of hornbeam at Anderson Pond an isolated and peculiar event, or was it part of a more widespread phenomenon? It was only after we and our students had completed the study of a series of sites from the Appalachian Mountains to the Ozark highlands and had prepared our time-series of maps for eastern deciduous forest species that the magnitude of the changes in hornbeam populations became apparent. What we found was that during the time of changeover from Pleistocene to Holocene climate, hornbeam became important throughout the Lower Midwest region, and then diminished, after about six thousand radiocarbon years ago, never to regain its former abundance. But rather than immigrating from somewhere to the south, hornbeam had been part of the late Pleistocene forest. It simply expanded in place. Often it was part of a floristics relay, a race among plant species to occupy space and gain access to sunlight. For example, at Cupola Pond, Missouri, hornbeam replaced black ash and was replaced in turn by hickory (Figure 3).

The expansion of hornbeam was thus a region-wide event, an example of an ephemeral plant community unlike any on today's landscape. This event coincided with the time in the most recent Milankovitch climate cycle during which seasonal contrast in solar radiation was at a maximum, that is, with very hot summers offset by very cold winters. Hornbeam is thus an example of species that are resilient in the face of late spring and early fall frosts, and other disturbances such as fire or wind damage that might eliminate competing species less able to sprout from its trunk base. During the late-glacial and early Holocene intervals, plant communities were continually changing as boreal species died back at middle latitudes and were replaced by a succession of invading temperate deciduous forest species. Hornbeam and other species were temporarily important players on this continually changing stage, as a series of unique and ephemeral deciduous forest communities formed and dissolved under the influence of highly seasonal climate of the early Holocene.

Rates of change in the composition of biological communities were greatest during the initiation of northward migrations, at the Pleistocene/Holocene transition, and in the past one thousand years during which Native Americans and European Americans have been most influential in altering the vegetation. As we now know, the changes during times of global climate change are dramatic and have far-reaching consequences for the fate of biological communities.

Prehistoric Native Americans as Agents of Ecological Change

Ecologists have long viewed Native Americans as "noble savages" living in harmony with their environment but having essentially no influence upon the distribution or composition of native vegetation. In a series of books, Professors William Cronon of Yale University, Michael Williams of Oxford University, G. Gordon Whitney of Harvard University, and Emily Russell of Rutgers University explain how this viewpoint developed. It is an outgrowth of descriptions made by 18th and 19th century pioneer settlers and land speculators who encountered a vast, unbroken expanse of forest that to them seemed to stretch endlessly from the Atlantic Coastal Plain to the Great Plains. The eastern deciduous forest was seen by these colonists as a limitless resource. The first wave of American pioneer settlement occurred along the Atlantic Seaboard from Georgia, the Carolinas, and Virginia north to New England. Clearing the mature forests of this landscape was at first a slow process, accomplished with hand tools one homestead at a time. Only after canal systems were built and railroad lines were in place was the frontier pushed back at an accelerated pace. With the establishment of the General Land Office Survey, much of the interior of North America was surveyed systematically in the early to mid-1800s, and prime lands were identified both for timber harvest and for conversion to agricultural land. Native American villages encountered by these colonists were small and the general landscape was not densely populated. Activities of indigenous peoples included the use of fire to clear the understory of forests but this practice was restricted primarily to the vicinity of their villages.

During the 1800s, however, explorers noted an abundance of archaeological evidence of past civilizations and began to realize that much larger aboriginal populations must have existed in the past than at the time of European American settlement. Archaeologists Dan and

Phyllis Morse reviewed the evidence of ancient American civilizations in the Lower Mississippi River Valley where numerous large burial mounds had been found. Pot-hunters had plundered many of the largest of the burial mounds for valuable artifacts. The systematic survey of Indian mounds funded by the US Bureau of Ethnology in the 1880s identified the locations of many more ancient ceremonial centers. One of the most important of these centers was in present-day East St. Louis, in a broad lowland called the American Bottom, located at the juncture of the Mississippi and Illinois rivers. At the time of the General Land Office Survey in the 1800s, wet prairie and woodland covered the nearly uninhabited American Bottom. Almost a thousand years before, however, the landscape had been a mosaic of villages and agricultural fields centered on a civic ceremonial center, the gateway city of Cahokia, which was built largely between AD 600 and 1200, then abandoned after AD 1250.

Both archaeological and paleoecological research have helped to interpret archaeological enigmas such as the formerly great city of Cahokia. By examining botanical and zoological remains associated with human artifacts, the role of prehistoric Native Americans has been redefined. They are no longer seen in a passive role as merely part of a natural setting, but are now appreciated as having been active players on the ecological stage of North America.

The first Americans, the Paleoindians, arrived in the New World during the late Pleistocene and were well established by thirteen thousand years ago. James Dixon of the Denver Museum of Natural History has traced the several routes by which Paleoindians traveled to North America. These routes included a land bridge across the Bering Straits known as "Beringia," the waters of the coastal Pacific Ocean, and even the pack ice of the North Atlantic Ocean. Genetic evidence from DNA samples of contemporary native peoples indicates that there were as many as five different source populations from eastern Asia, Siberia, and even western Europe. These cultures had different adaptations for surviving in Ice Age environments. Some bands of nomadic Paleoindians traveled south from the Bering Straits region through an ice-free corridor that opened across Alberta as the western Cordilleran and eastern Laurentide ice domes melted at the end of the Pleistocene.

Seafaring peoples from eastern Asia dispersed to North America through passages along the North Pacific sea ice margin, island hopping and navigating by means of skin boats.

Early Paleoindians used diverse tool kits and were versatile hunter-gatherers capable of adapting opportunistically to variations in the availability of raw material and food resources. Big-game hunting, which was part of Paleoindian lifeways, came to an end about ten thousand years ago with the extinctions of many species of megafauna. These extinctions of large mammals, including mammoth (*Mammuthus*) and mastodon, resulted in part from changes in environmental conditions. During the transition from late Pleistocene to Holocene climate, extreme winter cold and extreme summer warmth triggered a series of changes in the biota. As ecosystems were restructured, many of the large mammals upon which Paleoindians had depended became extinct. As proposed by Paul Martin, excessive hunting may have been a second cause of reduction in species numbers. Native Americans responded to the loss of the megafauna by changing their adaptive strategies. In order to survive the Pleistocene/Holocene transition, humans were forced to seek new solutions in how they sought food, where they settled, and how they crafted tools.

In eastern North America, humans adapted to new conditions in the early Holocene by diversifying their subsistence system and by developing interconnected macroband networks located along major waterways such as in the Lower Mississippi River Valley and at the Fall Line where the Appalachian Piedmont meets the Atlantic Coastal Plain, for instance along the Savannah River. By ten thousand years ago, Late Paleoindian and Early Archaic people practiced a logistically organized foraging strategy, based upon establishing a semi-permanent residential base and a seasonal round of activities involving use of specialized field camps. They began to exploit storable mast including hickory nuts, acorns, walnuts, and chestnuts, all produced by temperate deciduous trees that were becoming more abundant across eastern North America with the change to interglacial climates.

Late Paleoindian and Early Archaic people were more archaeologically visible than their predecessors — they left behind greater numbers of durable artifacts such as projectile points and nutting stones. Overall, however, human populations were still relatively small and widely dispersed, with broad buffer zones separating macrobands during most of each year. Their lifeways as foragers would have affected the vegetation and availability of wildlife locally around their settlements and base camps. But because of their small populations and their minimalist lifeways as hunter-gatherers, they probably had only a nominal impact on regional landscapes.

Through the Archaic cultural period, from about nine thousand five hundred to three thousand years ago, across eastern North America Native Americans had a diversified, mixed economy in which both wild and domesticated native plant foods became an important supplement to game and shellfish. The transition from foragers to farmers occurred over a long time span. The first domestication of native plant species, beginning by about five thousand to four thousand years ago, occurred primarily as a result of human disturbance of floodplain habitats. There, semi-permanent residential sites created openings in the vegetation that were colonized by weedy species, including the first domesticated strains of native squash (*Cucurbita pepo*), which were probably first grown for their oily seeds rather than their fleshy rinds. Other native plants that were cultivated included chenopod (*Chenopodium berlandieri*), whose seed coat thinned over time as people selected the most edible seed heads for broadcasting seed. Sumpweed or marsh elder (*Iva annua*) was yet another floodplain herb whose starchy seeds were eaten by Native Americans; the increasing size of sumpweed seeds in the archaeological record represents evidence of domestication.

In the process of becoming horticulturalists, Archaic Native Americans began to alter their ecological setting. Smithsonian Institution archaeologist Bruce Smith has used the term "domestilocalities" to describe Archaic period villages along the major river systems of the eastern woodlands. Smith envisioned that by their everyday activities, which included making local forest clearings, trampling the ground along footpaths, and enriching the fertility of the soil around refuse pits, Late Archaic peoples encouraged riparian weeds to thrive in their "door-yard gardens."

Beginning in the Late Archaic and Early Woodland cultural periods, between about four and three thousand years ago, Native Americans began to have more widespread and long-lasting impacts on their environment. With domestication of native plants, people began to have a positive effect on the overall diversity of species and habitats in the vicinity of their village sites. Villages along rivers, however, were not just sites for food plant domestication. They also were places where other useful weedy plants such as American cane, Indian hemp (*Apocynum cannabinum*), nettle (*Urtica dioica* and *Laportea*), and milkweed (*Asclepias*) grew. The fiber these plants produced was used to weave cane mats and to spin fine yarn. As riparian species spread into areas disturbed by humans, plant succession took on a new form — plant species that required full sunlight and disturbed ground, such as occurred on gravel bars and stream banks, now had abandoned Indian fields to invade.

By the Early Woodland cultural period, Native Americans also grew indigenous food plants in garden plots at the entrances to rock shelter sites across the region from the Ozark highlands to the Appalachian Mountains. By girdling the bark of trees and setting local fires, they opened up gaps in the forest canopy that allowed for a patchwork of forest succession to take place. As human-set fires periodically burned the upper slopes near the rock shelters, fire-adapted plant species were favored, and the entire ecological gradient between moist lower slopes and dry ridge tops was transformed. The activities of Woodland period Native Americans thus were intermediate levels of disturbance that enabled a wider variety of plant communities to coexist than would have been possible without the stimulating effects of fire. The human use of the torch profoundly transformed the character of the eastern deciduous forest beginning about three thousand years ago.

By AD 1000, many Native Americans lived in sedentary villages in an agricultural-based society. With the introduction of maize (*Zea mays*), people became increasingly dependent upon its cultivation to sustain their growing populations. They switched from an economy based upon a diverse diet to one largely relying on sustained annual production of relatively few food crops. Hernando de Soto and his conquistadors arrived on the North American continent in 1540, forever changing the face of the southeastern United States, as reviewed by Professor Charles Hudson of the University of Georgia. Long before the arrival of the Spaniards, however, Native Americans in many parts of the eastern woodlands had exceeded their resource base and major ceremonial centers, including Cahokia, had been abandoned. With the introduction of infectious diseases from Europe in the 1500s, the indigenous human population declined precipitously. Over the next several hundred years, semi-natural vegetation regenerated on abandoned Indian old fields. Thus, secondary forest appeared to pioneer settlers of the 1700s and 1800s as "virgin" forest, which they perceived as having been untouched by human influences.

In two separate regions, one based in the southern Appalachian Mountains and the other in the Mississippi River Valley, Paul and I have had the opportunity to see first-hand the evidence for the ecological impacts of prehistoric Native Americans. In a series of studies beginning with our arrival in Tennessee in 1978, we coordinated paleoecological studies with archaeological investigations in western North Carolina, eastern Tennessee, southeastern Kentucky, northeastern Arkansas, and southeastern Missouri. Taken together, these studies have allowed us to reconstruct changes in Holocene landscapes in areas directly affected by the activities of prehistoric Native Americans.

In 1979, as the Tennessee Valley Authority (TVA) was in the process of closing the Tellico Dam, resulting in the inundation of the Little Tennessee River Valley, Paul and I prepared to core the sediments of Tuskegee Pond. The paleoecological record from this site was to be studied by Patricia Cridlebaugh for her dissertation in the UT Department of Anthropology. Tuskegee Pond was named for a historic Cherokee village that was once located adjacent to the pond. Tuskegee was one of about sixty Native American villages that dotted the valley of the Little Tennessee River in the 1700s, the time of contact with European settlers, and was the birthplace of Sequoyah, the father of the written language of the Cherokee. The valley of the "Little T" was then the homeland of the Cherokee Nation and was thus of intense interest to UT archaeologists, including Dr. Jefferson Chapman, who directed the excavations of prehistoric Native American sites.

Located between Knoxville and the Great Smoky Mountains to the southeast, the Little T has its headwaters at an elevation of about four thousand five hundred feet and empties into the Tennessee River some eighty river miles downstream. In its natural state, the Little T was a relatively shallow river, with many shoals where freshwater mussels (Unionidae) abounded. The valley of the Little T is contained within narrow gorges in the mountains, but the river spreads out where it enters the gently rolling land of the broad expanse of the Ridge and Valley Province to the west. Through millions of years, the Little T has cut down ever deeper within this broad valley as it has eroded its way through sedimentary rocks of Cambrian and Ordovician age. At least nine sets of stream terraces, mapped by Paul and Jeff Chapman, have been left behind in the process, each abandoned as the river changed its course on its long journey. The oldest stream terraces are the highest benches stranded on today's landscape (Figure 30).

The Little Tennessee River began to lay down the deposits of its youngest terrace fifteen thousand years ago. Over time, silts and clays accumulated, with a fine layer added each time the river overflowed its banks. Human artifacts were buried within late Pleistocene and Holocene stream deposits. Over many generations, hearths, campsites, homes, and gardens of aboriginal people who lived on the stream banks were inundated by the river in flood stage. Through the millennia, numerous cycles of flooding and reoccupation of living areas have preserved an exquisite record of continuous human habitation of the river valley dating back to Paleoindian time some ten to twelve thousand years ago.

Pat Cridlebaugh used two kinds of fossil evidence to understand the long-term impacts of prehistoric human occupation along the Little T. The first kind of evidence was ethnobotanical — that is, plant remains recovered by water-sieving sediment samples from archaeological sites. Prehistoric Native Americans chose the kinds of wood for their campfires and they selected the seeds and nuts they used for food. The ethnobotanical record, consisting of macroscopic bits of charred wood, seeds, and nutshells preserved in prehistoric hearths, is direct evidence of human use of available plant resources. Cridlebaugh's second form of evidence for ecological impacts of prehistoric humans came from fossil pollen grains, spores, and microscopic charcoal particles preserved in sediments of small, natural ponds located both within the Little Tennessee River Valley and in the uplands at a distance from the locations of most of the archaeological sites. By comparing the ethno-botanical and paleoecological records from within the valley, Cridlebaugh could ascertain both direct effects of human selection of plant materials for use and the more indirect imprint of their activities on the land surrounding villages and ponds. Then by contrasting the paleoecological record from within the zone of maximum human impact along the valley with that from an upland environment beyond the area of intensive human occupation, she hoped to gain insight into the extent that the overall landscape had been transformed through time by human touch and torch.

The Icehouse Bottom archaeological site was a critical source of ethnobotanical data that was central to Cridlebaugh's research. Icehouse Bottom was a broad piece of Terrace 1 that lay adjacent to a high bedrock knoll called Rock Crusher Bluff. Named for the architectural remnants of a pioneer icehouse and milk shed, the farmstead at Icehouse Bottom was abandoned when it was bought by the Tennessee Valley Authority and scheduled for flooding beneath the Tellico Reservoir. Icehouse Bottom was then available, along with some two dozen other sites, to be excavated completely during the 1960s and 1970s by Chapman and his work crew of archaeology students. The six-meter wall of silty sediment represents the full sequence of Archaic period deposits of Terrace 1, exposed to view by the diligent work of a skillful backhoe operator. Chapman is fond of saying that Icehouse Bottom is like a layer cake of prehistory. Some thirty separate human occupation levels had been buried there by backflooding from the Little T between about nine thousand five hundred and four thousand years ago. This river deposit records the history of prehistoric human settlements from the time Archaic people first established their seasonal encampments on the river banks.

From Early Archaic time on, people built hearths on the flood-plain surface. These hearths were lined with clay brought in from the uplands, and some of the fire-reddened hearths still bear the impressions of the woven cane baskets used to carry the clay. Within the hearths are the charred remains of plants, including wood charcoal left from kindling used to stoke cooking fires. Hulls of the hickory nuts, walnuts, and acorns Archaic people gathered from trees in the surrounding forest are also abundant. Carbonized fruits and seeds of other native plants, both herbs like pokeweed (*Phytolacca americana*) and vines like grapes (*Vitis*), are also contained within the hearths. Pokeweed, whose tender young leaves provide an edible salad green but whose fruits are poisonous, probably grew in frequently disturbed areas along the floodplain, much as it does today. Grapevines grew at the forest edge and provided a sweet fruit. The types of wood charcoal in the earliest cooking hearths dating from the Archaic cultural period are from trees that would have grown near the Indian encampments. The floodplain forest included ash, elm, and red maple. Cridlebaugh suggested that Archaic people seemed to have picked up downed branches and cleared the bottomland forest in the immediate vicinity of their camps, a practice that would have required the least amount of work effort.

Animal bones are poorly preserved at Icehouse Bottom, as at similar sites elsewhere throughout the southeastern United States, because the warm, humid climate promotes leaching and bone decay. Among the most abundant faunal remains are bones of white-tailed deer (*Odocoileus virginianus*) and wild turkey. Evidence of hunting is well preserved in the array of stone projectile points. At Icehouse Bottom, there is evidence that Archaic people developed a sequence of stone points through time as hunting methods changed. This sequence of changing projectile point types is documented clearly at Icehouse Bottom because layers of sediment have covered each human occupation level, preserving each in an undisturbed condition through the millennia. Because of the discoveries at Icehouse Bottom, archaeologists can now accurately date any Archaic period "arrowhead" found lying at the surface of a freshly plowed field in the southeastern United States by comparing its size and shape with those preserved in radiocarbon-dated deposits from the Little T.

Woodland period sites along the Little T contain evidence of cultivated plants, including not only squash and gourd (*Lagenaria siceraria*) but also chenopod and sumpweed or marsh elder. The Mississippian cultural period was marked by development of large villages throughout the Little T. These villages included ceremonial centers built on raised earthen platforms and surrounded by palisade walls. Villages were interspersed with agricultural fields of maize, introduced from the American Southwest one thousand six hundred radiocarbon years ago.

By the Mississippian cultural period, the ethnobotanical wood

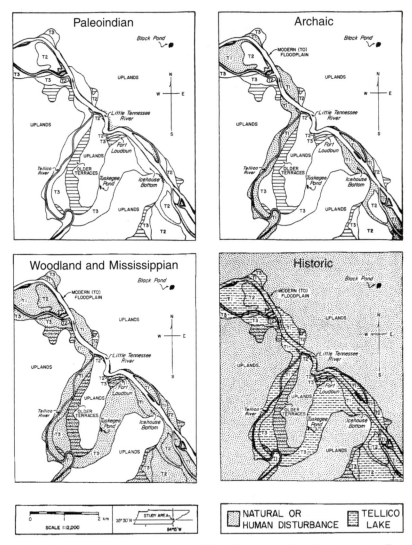

Figure 30. Changing landscapes of the Little Tennessee River Valley, eastern Tennessee. Stream terraces are designated by numbers; T1 is the late Pleistocene/Holocene terrace. Locations of Tuskegee Pond, Black Pond, Icehouse Bottom archaeological site, and historic Fort Loudoun are shown. Extent of land area affected by human activities is shown by shaded pattern. (Diagram from Delcourt and Delcourt, 1988)

charcoal record shows that wood burned in hearth fires included very little from bottomland trees, which were becoming scarce as more land was cleared, settled, and cultivated. Instead, pine, red cedar, and American cane became important in the wood charcoal record. Today, Virginia pine (*Pinus virginiana*) and red cedar grow in abundance on abandoned agricultural land throughout eastern Tennessee. By Mississippian time, these scrub trees had invaded Indian old fields. Split cane was used extensively for weaving baskets, mats, and even for thatch roofs. Cane is most abundant along river banks that have been cleared of trees. Today, rapidly expanding cane brakes are held in check only by cattle that eat the tender shoots back to the ground.

The Historic period began by the 1520s, when the first Spanish explorers traversed the southeastern United States. The Spaniards brought peaches (*Prunus persica*), which were brought into cultivation quickly by the Cherokee. When English settlers began to arrive in the 1700s, the people they met along the Little T were known as the "Overhill Cherokee" because they lived over the hill (the Great Smoky and Nantahala mountains) from other Cherokee populations in northwestern Georgia and the Carolina Piedmont. Some archaeologists think that the Cherokee were not the descendents of the Mississippian-age people of the Little T. Instead, it is likely that Mississippian peoples of the Little T were driven out and replaced by the aggressive Cherokee some time between five hundred and three hundred years ago.

The English established Fort Loudoun, the first British colonial fort west of the Appalachian Mountains, about one and a half kilometers from Icehouse Bottom in 1756. Relations between the British and the Cherokee soon deteriorated, and the Cherokee burned the fort to the ground in 1760. In 1762, British Lieutenant Henry Timberlake dew a map of the distribution of Cherokee towns throughout the Little T. This map became the basis for later excavations of the principal historic and prehistoric areas of human occupation.

Naturalist William Bartam explored the region on horseback in 1775 and 1776. Upon entering the Little Tennessee River Valley near its headwaters in the Nantahala Mountains of southwestern North Carolina, he described an enchanting scene where Cherokee maidens picked wild strawberries (*Fragaria virginiana*) in verdant fields, and where

185

wild turkey and deer bounded in the meadows. By the early 1800s, however, the situation was far from idyllic. The American government ordered the forced removal of the Cherokee from their homeland, and in 1838 they were marched overland to Indian territory in what is now Oklahoma. This devastating event, in which some four thousand Cherokee perished, is known as the "Trail of Tears." Only a few Cherokee escaped removal and remained in the southern Appalachian Mountains. Their descendents reestablished the eastern band of the Cherokee and today reside on the Qualla reservation bordering the southeastern margin of the Great Smoky Mountains National Park.

Paleoecological studies from Tuskegee and Black ponds helped Cridlebaugh to complete the portrait of long-term changes in the landscape mosaic of eastern Tennessee (Figure 30). Tuskegee Pond, only about an acre in size, was located on Terrace 3, at an elevation intermediate between the active stream channel and the hilly uplands. Tuskegee Pond was also within sight of Fort Loudoun, and thus recorded in its sediments both prehistoric and historic changes in the landscape of the central Little Tennessee River Valley. Black Pond was a small sinkhole pond in the uplands about four kilometers from Tuskegee pond. The combined paleoecological records from the two ponds span the past several thousand years of the Late Woodland, Mississippian, proto-Historic, and Historic cultural periods.

The pollen record from Tuskegee Pond shows that by Woodland time, people had converted low terraces along the Little T to a cultural landscape that was a mosaic of villages, footpaths, garden plots, secondary scrub pine stands, and cane brakes. Pollen of maize is consistently represented in sediment samples, indicating that fields of maize were cultivated on Terrace 3 near the pond. Ragweed (*Ambrosia*) makes up as much as sixty percent of the fossil pollen assemblage, and the high percentages of herbs were interpreted by Cridlebaugh as evidence for extensive land clearance on the stream terraces. Along with the evidence of Indian old fields in the Tuskegee Pond record, high accumulation rates of charcoal ash particles were found, indicating that Native Americans used fire not only for cooking and for warmth, but also to manage the secondary scrub vegetation on the terraces of the Little T. In sharp contrast, the pollen record from Black Pond for the

Woodland and Mississippian periods shows that regional uplands remained clothed in forest of oak, chestnut, hickory, and pine, with much less charcoal evidence for fire. Sediments from Black Pond contained less than two percent ragweed pollen and no maize pollen until about four hundred years ago, corresponding to the time of first European contact.

During the Holocene, the Little Tennessee River Valley has undergone a dramatic transformation from a natural to a culturally modified landscape (Figure 30). This process was underway long before modern Europeans devised a means to sail across the Atlantic Ocean. Cridlebaugh's study demonstrated that prehistoric Native Americans did not live in total harmony with their environment, in contrast to prevailing notions held even by early naturalists such as Bartram. Instead, they lived within the context of a changing landscape that was being altered gradually by their own actions.

In the Woodland cultural period, Native Americans cultivated food plants along the major river valleys in the eastern woodlands. They also actively occupied rock shelters on bluffs overlooking river valleys, using these sites as dwellings and hunting camps. Charred remains of cultivated food plants are abundant in ancient hearths found within rock shelters of both the Ozark highlands and the southern Appalachian Mountains, dating to the time of maximum use of rock shelters in Late Archaic and Early Woodland times. To what extent did these prehistoric Native Americans influence the vegetation on surrounding hill slopes through use of fire and establishment of garden plots?

To find a suitable study site with which to answer this question, we sought the help of Rex Mann, Cecil Ison, and Bill Sharp of the Daniel Boone National Forest, located in the heartland of E. Lucy Braun's mixed mesophytic forest region in eastern Kentucky. In 1996 we cored the sediments of a woodland hollow pool situated on a high ridge top along Keener Point, near Berea in eastern Kentucky. We called the site Cliff Palace Pond because of its location within a few tens of meters of a large, amphitheater-like rock shelter that had been dubbed Cliff Palace by Forest Service archaeologists. The pond is nestled in a small depression within sandstone, with a nearly continuous canopy of red maple and sweetgum trees overarching it. No more than one hundred fifty square meters in area, Cliff Palace Pond is only about half a meter deep at its center. Its setting imparts a mystical quality to this site, because it is linked with the Cliff Palace rock shelter by a flight of ancient stairsteps carved into the bedrock. The rock shelter contains abundant evidence of prehistoric inhabitation, including nutting stones used to shell acorns and hickory nuts and a spiral petroglyph, representing a fertility symbol, etched in a fallen slab of sandstone.

We knew from radiocarbon dates of charred food remains in hearths from this and other nearby rock shelters studied by Ison and Kristen Gremillion of Ohio State University that the time of their most frequent use was during the Early Woodland cultural period, beginning about three thousand years ago. Charred seeds of native cultivated plants of the "Eastern Agricultural Complex," including sunflower (*Helianthus annuus*), marsh elder, chenopod, little barley (*Hordeum pusillum*), maygrass (*Phalaris caroliniana*), and even ragweed are typically found in Woodland-age hearths within rock shelters throughout the cliff section of eastern Kentucky. This assemblage of food plants, however, was incongruous with the notion that mixed mesophytic forest blanketed the region with a continuous forest cover of hardwood trees. E. Lucy Braun described the Cliff Section of the western Cumberland and Allegheny plateaus as primarily mixed mesophytic forest on slopes lying below the sandstone cliffs. The cliffs themselves supported oak-chestnut forest, with pitch pine (*Pinus rigida*) clinging to the driest outcrops and an open understory of bracken fern, little bluestem grass (*Andropogon scoparius*), lichen, and moss. With paleoecological evidence from Cliff Palace Pond, we hoped to explain this mixture of plant communities on Keener Ridge.

From the fossil pollen evidence, we found that before three thousand years ago the vegetation was composed mainly of fire-intolerant hardwoods such as American beech, sugar maple, and eastern hemlock (Figure 31). With human occupation of nearby rock shelters beginning three thousand years ago, however, a dramatic change took place in the composition of forests in the vicinity of Keener Point. Oak and American chestnut became the dominant trees on upper slopes and fire-adapted pitch pine established on the ridge crest. Large increases in accumulation rates of charcoal particles, traces of pollen of sunflower, chenopod, marsh elder, and ragweed, and fragments of fire-cracked rock in sediments of Cliff Palace Pond reinforced our interpretation that human intervention was responsible for these changes in forest composition.

Woodland-age peoples evidently used fire not only for cooking and warmth, but also to clear the upper slopes of ridges for garden plots. Archaeologist Wesley Cowan estimated that a clearing of four hundred square meters would have been sufficient to grow enough chenopod plants to feed a family for several months during the late-winter time of low food resources. This size of garden clearing is similar to that made by girdling and felling several canopy trees. Light gaps thus created by Native Americans would have been sites for forest succession after shallow soils were depleted of nutrients by only a few years of gardening. In the paleoecological record, we found pollen of successional hardwood trees, such as tulip tree, which require high light intensity to germinate and grow through seedling to sapling stage in a forest environment.

The paleoecological and archaeological evidence from Cliff Palace Pond and Rock Shelter led us to conclude that the high diversity of plant communities within E. Lucy Braun's beloved mixed mesophytic forest was augmented and maintained by traditional activities of prehistoric Native Americans. During much of the 20th century, with a policy of fire suppression maintained by the United States Forest Service, neither oak nor pitch pine regenerated well on Keener Point or elsewhere in eastern Kentucky. Consequently, fire ecologist Rex Mann and archaeologist Cecil Ison are now using selective, prescribed burning in a way that mimics the traditional practices of Native Americans to re-introduce fire as an ecological factor in the mixed mesophytic forest region of eastern Kentucky.

As reviewed by Professor Charles Redman of Arizona State University, in Late Woodland and Emergent Mississippian times, when people became tethered to fixed sites and when population densities grew to more than fifty people per square kilometer of cultivated land, environmental degradation often followed as ecosystems were fragmented by ever-expanding agricultural fields. We had found one example of over-exploitation of resources in the Little Tennessee River

Valley, where Mississippian peoples were replaced in the Proto-His-toric period by the Overhill Cherokee. In other areas, such as in the center of the Mississippian cultural influence at Cahokia, the ecological crisis resulting from overpopulation combined with climatic variability led to cultural instability and eventual evacuation of the ceremonial center. Working together with Dan and Phyllis Morse of the Arkansas Archaeological Survey and Roger Saucier of the United States Army Corps of Engineers, in 1999 we mapped Holocene changes in bottomland vegetation in the central portion of the Mississippi River Valley (Figure 32). The maps were based upon Saucier's geological mapping of riverine habitats, eleven paleoecological sites, and the Morse's extensive collections and mapping of archaeological finds throughout northeastern Arkansas. By collating these several data sources, we were interested in finding out when the significant landscape threshold was reached that both fragmented the bottomland forests and preconditioned the demise of Mississippian cultures.

Dan and Phyllis' work had shown that ten thousand years ago, during late Paleoindian time, small bands of Dalton people had occupied the riverine corridor, which was largely forested with bald cypress-tupelo gum in permanently wet backswamps, sweetgum-ash forest on levee slopes that remained inundated through much of the year, and oak-hickory forest on drier levee crests. Dalton people relied on a variety of small game, fish, and plant resources and used woodworking tools such as adzes to carve whole bald cypress trees into long canoes for travel along the meandering Holocene Mississippi River. Dalton people clustered as small bands in different local watersheds and were probably present in such small numbers that they had little impact upon native plant and animal resources.

Through the Archaic and Woodland cultural periods, climatic warming and drying that characterized the mid-Holocene through much of the midwestern United States was evidenced in the Central Mississippi Valley by lowered water tables and a change in vegetation to open savanna on highest floodplain and braided-stream terrace surfaces, with bottomland trees confined to the levees and backswamps of the active Mississippi River meander train. Only after two thousand years ago did increases in precipitation result in rising groundwater, increased

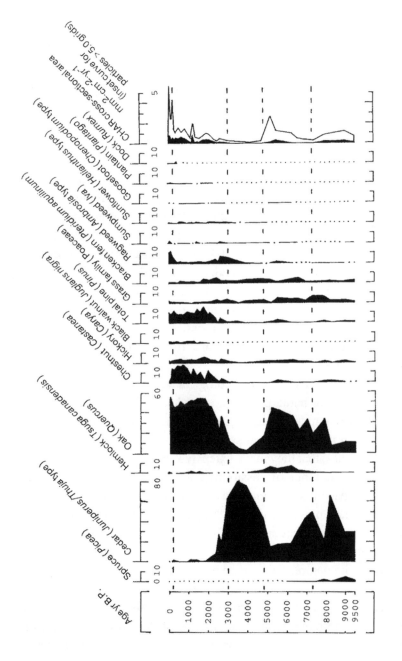

Figure 31. Summary diagram of pollen and charcoal from Cliff Palace Pond, Kentucky. (Diagram from Delcourt *et al.*, 1998)

191

overbank flooding, and expansion of sweetgum-elm forest. The environmental changes during the mid-Holocene may have resulted in changes in adaptive strategies of Archaic people. Little archaeological evidence is available for Archaic people in the Mississippi Valley of northeastern Arkansas. As previously navigable streams dried up into ephemeral sluiceways, people may have moved from unproductive bottomland to adjacent uplands where game and mast would have increased because of increased forest edge in open savannas. With a late Holocene shift to increased water flow in the Mississippi River and its tributaries, bottomlands once again would have become hospitable, and Woodland peoples would have found them an attractive setting for villages during the time of transition from foraging to farming.

Widespread deforestation and conversion of bottomlands to agricultural land began around AD 700 in the Mississippian cultural period. The paleoenvironmental map for one thousand radiocarbon years ago (Figure 32) shows that by the height of the Mississippian cultural period, prehistoric Native Americans had completely fragmented the natural forest vegetation into a patchwork of forests and fields, not only on active floodplains of the Mississippi and its tributaries but also on braided-stream terraces at a distance from the active channel of the Mississippi River.

Recent synthesis of paleoecological and archaeological research is leading to a new paradigm. We now understand that for a given place and time, we cannot assume that Native Americans had no discernable influence on their ecological setting. Nor can we conclude, without specific studies, that they made extensive alterations to the natural environment. The range of possibilities is broad, and examples can be found that span the spectrum. One conclusion is evident — the concept of a Holocene, "presettlement" or "virgin" forest, untouched by human hands, is a myth. As concluded by Professor Shepard Krech III of Brown University, Native Americans neither existed apart from nor in harmony with nature. This notion was based on false impressions and stereotypes first developed by European American colonists. To the contrary, prehistoric human activities in many cases had long-term consequences, beginning with the extinction of Pleistocene mammals at the end of the Pleistocene, and continuing through the domestication of

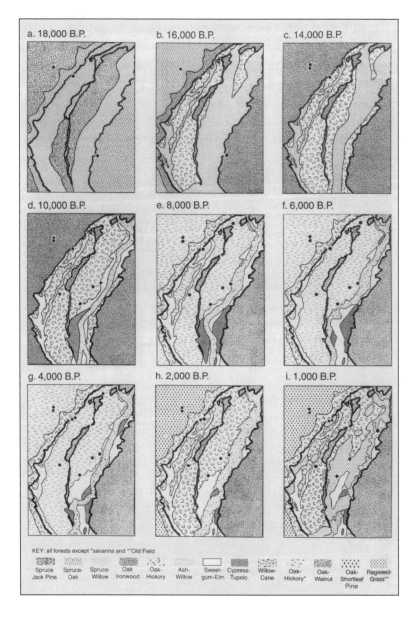

Figure 32. Mapped changes in Holocene landscapes of the central portion of the Mississippi River Valley, showing the two times of greatest change in prehistoric times — extinction of Critchfield spruce and establishment of warm-temperate bottomland trees by ten thousand years ago and conversion of bottomland forest to agricultural land one thousand years ago. (Diagram from Delcourt *et al.*, 1999)

193

plants still in use today, such as squash, and through landscape modifications made by land clearance, intensive maize agriculture, and selective use of fire. During the Holocene, across North America there was a cultural mosaic of human impact along gradients in space and through time. Human cultures evolved as part of nature, not apart from it, but humans are a unique ecological factor in having both cultural interactions and accumulated knowledge based on the experience of past generations.

Lessons from the Past, Implications for the Future

Earth is now at a crossroads for climate change. According to calculations based on the Milankovitch astronomical theory of the progression of ice ages, the present interglacial interval should be nearly over. Global climate should begin to cool within the next several thousand years, leading to growth of glacial ice and the onset of the next Ice Age. John Imbrie was among the first to suggest that humans are having an unprecedented effect upon Earth's climate that will override the natural pattern of variation in solar insolation. Since the Industrial Revolution, the concentration of carbon dioxide (CO_2) and other "Greenhouse gases" in the atmosphere has increased dramatically. The resulting warming effect is projected to sweep us all into what Imbrie called a "super-interglacial" in which global temperature will rise rapidly to levels well beyond those documented for any previous interglacial interval of the Quaternary. Can we use the lessons of the past to project the consequences of Greenhouse warming for the future of the eastern deciduous forest?

Since the Industrial Revolution, enormous amounts of carbon have been released into the atmosphere in the form of CO_2 gas. The level of atmospheric CO_2 has risen from 280 parts per million in pre-industrial times, about 250 years ago, to more than 360 parts per million today. The relationships between increases in Greenhouse gases in the atmosphere and climate change are the basis for computer models that simulate the future of the earth's climate system. For the 21st century, given a doubling of atmospheric CO_2 from pre-industrial levels, most models predict that average global temperature will be 1.5 to 4.5 ° Celsius warmer than in the 20th century. This temperature increase will be accompanied by regional shifts in weather patterns. Climate modelers estimate that in the next twenty-five years, global and regional climates are going to change more than they have in at least the past century.

Typical global temperatures will be higher in the next hundred years than at any time in the past ten thousand years of the Holocene interglacial. In the next several centuries, temperatures will exceed those of the last *one hundred million years.*

Climatologists Kwang-Y. Kim and Tom Crowley have projected trends in the earth's average temperature over the next one thousand to ten thousand years, presuming that humans will use a substantial amount of existing fossil fuel reserves and that the resulting input of CO_2 into the atmosphere will be essentially uncontrolled. Crowley and Kim forecast a spike in overall temperature between the years 2200 and 2400 that will probably reach 4 to 13° Celsius warmer than today, with warming of as much as 2 to 5° Celsius persisting as long as ten thousand years into the future. For context, over the past twenty thousand years, the average temperature of the globe warmed by only 5° Celsius between the extreme cold conditions of the Pleistocene and the modern interglacial climate. Only a few centuries into the future, the earth may experience "super-interglacial" warming exceeding that occurring any time since the Cretaceous Period. Furthermore, according to Kim and Crowley, human-induced climate warming will stack upon orbital changes during the next phase of the Milankovitch astronomical cycle, causing elevated temperatures to persist for as much as the next ten thousand years.

M. F. Loutre and A. Berger of the Université Catholique de Louvain, Belgium, consider that global climate warming will have even more enduring effects, in part because future changes in solar insolation will be less dramatic than those in many previous interglacials. They suggest that Earth's climate system will stabilize in this anthropogenic regime and not be overridden by natural Milankovitch cycles — the current interglacial interval could last another *fifty thousand years.* Antarctic and Greenland ice sheets would shrink and persist at minimal volumes until CO_2 levels fall below the threshold for ice sheets to grow once again. The next glacial maximum may not be reached until one hundred thousand years after present. This means that the remainder of the present interglacial will not only be much warmer than previous ones, but that it will be much longer in duration than any since that of "marine isotope stage 11" some four hundred thousand years ago. It now seems inevitable that the overprint of human-induced climate change will make the Holocene an extended time interval of global warmth without precedent in Quaternary history. Rapid, large, and long-lasting increases in global temperature will have significant effects on the biosphere. Future climate change will place unprecedented stress on many species of plants and animals.

What will be the consequences for species of the temperate eastern deciduous forest? Temperate forest species are already imperiled because human land use has become increasingly intense during the late Holocene. Forest communities that once occupied large and nearly contiguous sectors of eastern North America are today fragmented into small, discontinuous forest patches in a landscape that in only the past few hundred years has been converted in large part to agriculture, suburban, or urban land use.

My personal and professional odyssey as a historian of deciduous trees has brought me to the realization that the future of the eastern deciduous forest is now at risk. What have Quaternary paleoecologists learned about forest history that can aid in understanding how great that risk is? How can we use our knowledge of the past to help preserve biological diversity into the future?

In this book, I have identified several major ways in which ideas about the nature of vegetation have changed during the latter part of the 20th century. In large part, these changes in perspective have come about because of insights gained from fossil pollen studies. Let's consider each of these lessons from the past and their implications for the future.

We now realize that the evolution of modern species and the development of present-day biomes have taken place within the context of long-term tectonic and climate changes, and that variability in climate is a fundamental and ongoing natural process that affects the entire biosphere. Certain vegetation types that are now geographically extensive, for example arctic tundra and temperate grassland, are composed of species that evolved and assembled into communities largely within the latest Tertiary and Quaternary periods of cooling global climate. Other vegetation types, including temperate deciduous forest, are today globally more restricted in distribution than they were during the Tertiary Period because of Quaternary tectonic events such as uplift of the Alps and Ice Age climate cooling that disrupted their previous contiguity. During the twenty glacial-interglacial cycles of the Quaternary, the areas of temperate deciduous forest and of other major vegetation types have fluctuated. Glacial-age times of stability and vegetation unlike that of today characterized eighty to ninety percent of the time and were followed by rapid change to interglacial conditions and major reorganizations of biotic communities. Typically, interglacial intervals have been times of continual adjustments in the biota until the onset of the next glacial cycle.

On a time scale of hundreds to thousands of years vegetation is dynamic rather than static. The dynamic nature of plant communities is exemplified by the large changes in distribution and composition of vegetation during the past twenty thousand years. For some forest communities that existed south of the glacial margin during the late Pleistocene, we can find similarities in composition with the modern vegetation, but for many past communities there are no close modern analogs. For instance, jack pine and spruce forest that was widespread across the southeastern United States during the late Pleistocene resembled today's boreal forest only superficially. One significant difference was the presence of a now-extinct species of spruce whose ecology is largely unknown. In the Blufflands bordering the Mississippi River Valley, Critchfield spruce and temperate hardwoods were intermingled on windswept, silt-covered hill slopes where summers were cooled by advection fog. This now-extinct ecosystem was different from any on the modern landscape.

Subtle differences probably also existed in the structure and density of the forest stands. Jack pine forest in middle Tennessee grew on loamy, calcium-rich soil that was quite different from the sandy glacial outwash plains of the northern Great Lakes region where jack pine has been abundant in the late Holocene. Black ash and hornbeam were common in full-glacial boreal-like forests, allowed to intermingle with spruce because low winter temperatures were not as extreme as in the Holocene. Finally, across much of the southeastern United States, full-glacial climates were more arid than today, as indicated, for example, by active sand dune formation on the Atlantic Coastal Plain of the

Carolinas. Consequently, the jack pine stands of the late Pleistocene probably grew in open stands rather than as closed forest.

South of the glacial margin in eastern North America, the vegetation patterns of eighteen to twenty thousand years ago had reached equilibrium within a relatively stable, equable, though colder climate than today's. But once the Laurentide Ice Sheet began to retreat, and across the North Atlantic Ocean the sea-ice cover began to melt, an extended time of climate instability and biological reorganization ensued. Within hundreds to only a few thousand years, climate south of the glacial margin became unfavorable for tundra and boreal forest species, causing them to die back at the southern margins of their ranges and allowing for other plant communities to emerge.

The long-term dynamics of vegetation can be seen in the changes in the sharpness of ecological transition zones, or ecotones, between plant communities through time. During the full-glacial interval, an abrupt ecotone between boreal forest and deciduous forest was anchored at 34° N latitude. Defined by the changeover from jack pine and spruce forest to oak forest, this ecotone shifted northward after about sixteen thousand five hundred radiocarbon years ago as boreal trees died back and were replaced by advancing populations of oaks and other deciduous trees. Soon after that time, however, the ecotone broadened and became less sharply defined as the eastern deciduous forest began to diversify into a larger number of distinct, though rapidly changing, communities. Several additional ecotones developed, separating deciduous forest communities from one another. After ten thousand radiocarbon years ago, on the western margin of the deciduous forest region, a new ecotone formed between forest and grassland, as increasing seasonal contrast led to development of an extensive prairie region. During the early to middle Holocene, prairie plants that survived the late Pleistocene as local populations within forest openings expanded to form extensive grassland across the Great Plains. These examples illustrate that plant communities are not "rooted in place," catatonic in the face of environmental change. On the time scale of the late Quaternary, plants have formed dynamically changing communities because of the abilities of plant species to spread their propagules to new and favorable locations and to adjust their ranges in response to environmental change.

Another discovery based upon fossil pollen studies is that many species are resilient in the face of climate change and they respond differently according to their individual tolerances and life-history strategies. This means that the dynamics of vegetation change are not a matter of wholesale displacement or compression of intact life zones. The individualistic nature of species responses leads to novel and unpredictable combinations of communities that change continuously through time. Thus, some biotic communities are ephemeral and do not consist of tightly coevolved species dependent upon one another for survival. Margaret Davis was the first to demonstrate that both the directions and rates of postglacial tree migrations were individualistic. Davis's work further shows that the probability of long-term survival of species is conditioned by the time they need to adjust their ranges. Delays, or lags, in dispersal and establishment of new populations make for differential survival during times of rapid climate change. For example, during the late-glacial interval, oak and hickory, with their refuges generally south and west of the southern Appalachian Mountains, spread from south to north as a latitudinal wave. American beech, with its primary refuge in the Blufflands, migrated from southwest to northeast generally west of the Appalachians into the Great Lakes region. Eastern hemlock and white pine moved out of refuges on the continental shelf off of the Carolinas and spread westward and northward along the Appalachian Mountain chain, and then westward across the Great Lakes region. American chestnut was the last of the migrant tree species, moving out of the Gulf Coastal Plain only after about nine thousand years ago and spreading slowly northeastward along the Appalachian Mountains. American chestnut reached its northernmost limit of distribution in Connecticut only two thousand radiocarbon years before present, just a few generations before the chestnut blight was introduced. The differential directions and rates of tree migrations during the Holocene illustrate that not only do plant species differ in their tolerances and dispersal abilities, but also that their offspring require sufficient time to germinate from seed, grow, and reproduce in order for the whole population to adjust to environmental change.

Many studies in Quaternary paleoecology have shown that species can survive restriction within small, relatively isolated refuge areas and they can spread outward to new locations during times of changing climate — if they have access to continuous migration corridors. Long-term survival of species is, however, dependent upon overcoming bottlenecks for migration in the form of unfavorable climate or soil or other discontinuities in habitat. The Blufflands served as a significant corridor for species migrations during the late Quaternary. This region presented a relatively continuous set of habitats that allowed species to adjust their ranges during times of changing climate.

The potential threat to species posed by bottlenecks for migration is exemplified by the crisis experienced by boreal forest species during the late-glacial interval. With climate warming and an increase in seasonal temperature extremes, Critchfield spruce became globally extinct. Other spruce species and jack pine died out in the southeastern United States as climate warmed beyond their tolerance limits. Unlike Critchfield spruce, however, these and many other boreal forest species kept pace with the retreating Laurentide Ice Sheet. Both black and white spruce trees grew right up to the limit of ice as it melted back across the Great Lakes region. But just at the transition from late Pleistocene to Holocene conditions, ten thousand years ago, boreal forest species encountered a bottleneck that threatened their wholesale extinction. At that time, climate was warming and seasonal contrast was increasing faster in the southern portions of their ranges than ice was retreating to the north. They were literally caught between the summer heat and the ice wall. With further thinning of the ice dome, the Laurentide Ice Sheet eventually disintegrated and opened up vast new habitat in Canada, allowing tundra and boreal forest to expand onto newly deglaciated landscapes during the middle and late Holocene, as described by Canadian Professor E. C. Pielou. This lesson from the past illustrates that species can be resilient even during extreme environmental change. They become vulnerable to extinction, however, if their habitat is permanently reduced beyond that required for their populations to reproduce and maintain genetic viability.

Throughout the Holocene interglacial interval, humans have been an increasingly significant factor in shaping the destiny of the eastern deciduous forest. From the time that hunter-foragers increased enough in population size to become archaeologically visible some ten thousand years ago, their activities have had myriad impacts upon the biological communities within which they lived. Human disturbances on a local scale served to enhance biological and landscape diversity during much of the Holocene. Protection of nut trees in "sacred groves" and selective use of fire as a management tool would have promoted expansion of the "fagaceous" eastern deciduous forest. But even prehistoric human disturbance sometimes caused deleterious changes in landscapes, including deforestation, soil erosion, and overexploitation of plant and animal resources. Through most of prehistory, however, human activities were a kind of "intermediate disturbance regime" that affected the openness of vegetation to invasion by potentially weedy species. In eastern North America, old-field communities originated in Late Archaic and Woodland times as riparian plants colonized openings near villages located in bottomland deciduous forest. As certain Native American ceremonial centers, for example, Cahokia, were abandoned between AD 1250 and 1450, forest stands regenerated on many former village sites. Pioneer settlers of the 1700s and 1800s gained the false impression that the eastern deciduous forest had always been a wilderness with nearly contiguous late-successional vegetation throughout its extent.

Changes in the vegetation mosaic across eastern North America in historic times have resulted in the fragmentation of ecosystems, leading to a landscape with less of the core habitat that is critical for species requiring forest interior to survive. This newly opened landscape is now vulnerable to invasions by exotic species introduced from other continents. In its present configuration, the highly fragmented eastern deciduous forest no longer provides extensive continuous corridors for native plant migrations.

Taken together, these several lessons from the past have important implications for the future. The Quaternary vegetation history of eastern North America gives a context within which to evaluate potential future consequences of changes in climate. We can expect that with continued global warming, once again the deck of ecological cards will be shuffled. Climate change, outbreaks of pathogens, continued habitat degradation and fragmentation with increasing land use, and atmospheric pollution in the form of Greenhouse gases and acid precipitation all will continue to place stress on species survival. The outcome of this latest "natural experiment" cannot be known with certainty but potential shifts in species ranges and overall changes in the distribution and composition of communities can be projected given knowledge of their current tolerance ranges and their Quaternary history. In general, broadly tolerant "generalist" species will be favored over rare, narrowly endemic, or otherwise "specialist" species. Existing biological communities will be pulled apart, disassembled into their individual species components. Species that are common today may become uncommon or may be lost forever to extinction. New combinations of species will emerge that may include many species not originally native to North America.

What are the specific consequences of global Greenhouse warming for today's dominant plant species of the eastern deciduous forest? Margaret Davis and Catherine Zabinski of the University of Minnesota have prepared a series of maps depicting potential displacements in the optimal climate zone for several important species of the eastern deciduous forest, given Greenhouse warming that is projected for the 21st century. Their maps show large northward and elevational displacements for many of Braun's mixed mesophytic forest species, which will be forced to retreat from their current centers of distribution on the Allegheny and Cumberland plateaus and will potentially find refuge within the central and northern Appalachian Mountains as well as northward into Canada. Davis and Zabinski's maps depict the optimal climate conditions for American beech and sugar maple as located in northern New England and southern Labrador. Much of the potential future range for these species, however, is over poor substrates, such as glacially scoured rocks of the Canadian Shield that may prove to have unsuitable soil conditions for establishment of beech-maple forests. The rate and distance of range displacements projected for these species also outstrip their natural rates of dispersal and establishment based on the late Quaternary paleoecological record. Furthermore, the territory across which the species must migrate is largely a cultural landscape today. Agricultural, suburban, and urban land occupy large expanses across which the species must disperse if they are to be successful in surviving the coming centuries of global climate warming.

Researchers from the Northeastern Research Station of the USDA Forest Service have taken the preliminary findings of Davis and Zabinski to heart. A forest service team led by Louis Iverson developed an atlas in which they described the current geographic importance, relationship to environmental variables, and life history characteristics for eighty common tree species of the eastern United States. Iverson and Anantha Prasad developed a statistical model to project the changes in distributions of the species that could occur with global warming. They found that dramatic shifts in both centers of abundance and range margins are likely to occur as regional climate change restricts the optimal areas for many eastern deciduous forest trees to higher elevations in the Appalachian Mountains and to northern latitudes. By taking into account the patchwork nature of forested habitats, Iverson and colleagues are providing models of individualistic tree migrations that potentially could occur in the next hundred years across landscapes fragmented by human land use. From these models, Iverson and Prasad can estimate the probability of success or failure of different species to survive into the future based on their individual abilities to cross barriers to migration and the time lags for response to climate change inherent in their dispersal and growth characteristics.

Researchers such as Robert Thompson of the United States Geological Survey are also approaching these questions using a series of mapped overlays of present patterns of climate, hydrology, and vegetation together with global and regional models of climate change to project integrated landscape-level ecological shifts over the next century. Results of a related "VEMAP" vegetation mapping and forecasting project indicate that western and southern portions of the present eastern deciduous forest region will become increasingly droughty as well as warmer. Decreased soil moisture will turn much of the present deciduous forest region into a sparse woodland or shrubland that will become increasingly vulnerable to northward expansion of coastal plain pines as well as to invasion by exotic species. Enclaves of habitat suitable for mesic deciduous forest will eventually become restricted to favorable sites in the Appalachian Mountains, the margins of the upper Great Lakes, and higher elevations in New England. Taken together, the individual species models of Iverson and the habitat shift models of VEMAP are the kind of tools that conservationists and land managers need to estimate the potential changes in the eastern deciduous forest that will result from continuing global warming and to develop realistic plans for future ecological sustainability.

The Appalachian Mountains are a significant "hot spot" for biological diversity within the eastern deciduous forest region and will be a crucial region for future persistence of eastern deciduous forest species. The late Quaternary history of this region serves as a prologue for future changes and its study generates insights that conservationists need to address in order to preserve biological diversity into the future. Along the Appalachian Mountains the wide diversity of habitats presented by the broad gradient in elevation allowed for shifts in ecotones between alpine, boreal, and temperate plant communities, leading to the modern-day diversity of plant life and life zones and providing a basis for long-term survival of species. For example, the ecotone between red spruce-Fraser fir and deciduous forest, defined by the abundance of pollen of *Picea* and *Abies* relative to the pollen of deciduous trees detected in the fossil pollen record, shifted dramatically during

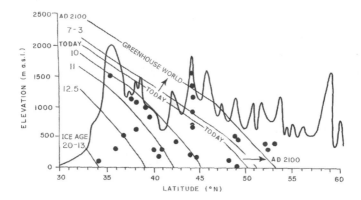

Figure 33. Changing position of the ecological transition zone between spruce-fir forest and deciduous forest in the Appalachian Mountains from twenty thousand years ago (20 x 10³ yr) to today and predicted future position of the spruce-fir/ deciduous forest ecotone with global warming. Locations of plant fossil sites used to document past ecotone shifts are shown as black dots. The topographic profile represents elevations of the highest Appalachian summits. Future altitudinal shifts are based on lapse rates assuming a 3° Celsius increase in mean July temperature. (Diagram modified from Delcourt and Delcourt, 1996)

the transition from Pleistocene to Holocene conditions (Figure 33). Spruce-fir forest covered most of the landscape of the southern Appalachian Mountains at latitude 35°N to an elevation below five hundred meters during the full-glacial interval. Between thirteen and seven thousand years ago, spruce-fir forest retreated above fifteen hundred meters and moved northward more than twenty degrees of latitude into the maritime provinces of eastern Canada, which left local populations only on the highest peaks in the southern Appalachian Mountains. In the Great Smoky Mountains, today the species richness of the flora decreases overall with increasing elevation and with decreasing area of mountainside available as habitat on the highest summits. Paradoxically, the number of rare and endangered vascular plant species increases markedly with increasing elevation. Half of these rare plant species found at high elevations grow in patchy open-ground habitats such as wet meadows, landslide scars, and cliff faces. Dependent today on local site conditions for their continued persistence, these relicts of a more widespread Pleistocene alpine flora are particularly vulnerable to extinction in the next century as global and regional temperatures rise. Projections of future changes in vegetation of the Great Smoky Mountains (figures 33, 34) based on conservative models of global warming indicate that red spruce-Fraser fir forest with its alpine species will be eliminated locally. In its place, deciduous forest will regenerate, with relatively cold-hardy species such as yellow birch, American beech, sugar maple, and mountain ash (*Sorbus americana*) predominating on mountain crests. Mixed mesophytic forest may expand from low-elevation coves and stream valleys to occupy mesic mid-slopes, with mixed oak forest on drier sites. Fire-adapted pine and heath communities will grow on the most exposed ridges at mid-elevations. Warm-temperate southern mixed hardwoods and southern pine forest will blanket low-elevation valleys.

In the Appalachian Mountains, over the past twenty thousand years, environmental changes have affected biological diversity at every level of biological organization, from genetic differentiation within species to gradients in composition of biological communities and landscape-scale pattern and process. In considering the future of biological diversity, several alternative approaches to conservation are possible.

We can take measures to preserve the present-day habitats for individual species that are rare, endemic, or threatened by habitat destruction. By taking this approach, however, we may ignore the widespread environmental changes that may compound more local threats to continued existence of species that are narrowly adapted to specific habitats. Alternatively, we can prepare for environmental changes that may make present-day habitats no longer suitable for rare species, focusing our attention on understanding the underlying processes of change and species adaptations to environment. If we can predict where the loss of biological diversity will be the greatest, we can provide corridors to allow for species to migrate successfully in the face of climate change. We may also need to be prepared to transplant endangered species to new locations where climate will be favorable. In so doing, we cannot ignore those species that are common today but may be rare in the near future because of climate change.

Within the Appalachian Mountains one of the biggest challenges for conservation of eastern deciduous forest species will be to either provide a massive planting program that will hand-carry seedlings to favorable sites, or to provide for habitat continuity that will allow the species to adjust their ranges by natural processes. For example, maintaining a nearly contiguous series of biological reserves in the form of state and national parks and state and national forest lands would allow for dispersal of species and minimize bottlenecks to migration of plants. Even in the Appalachian Mountains, however, there is neither the altitudinal nor the latitudinal extent to insure long-term survival of all currently imperiled species. Greenhouse world climate projections indicate that even in the northern Appalachians the ecotone between alpine and boreal ecosystems will be eliminated by the year 2070.

Arctic, boreal, and cool-temperate species may be affected the most by future climate change. Plants that evolved during the late Tertiary and Quaternary in response to the trend toward colder climates and permafrost development at high northern latitudes will be vulnerable to extinction as the permafrost thaws over widespread regions of northern Alaska, Canada, and Russia. On the other hand, warm-temperate and tropical species may be much less prone to extinction with future global warming. As summarized by Professor Zack Murrell of

Appalachian State University and colleagues studying floristic botany throughout the southeastern United States, across the southern Atlantic and Gulf coastal plains hundreds of vascular plant species are documented to exist in small local populations in geographically restricted areas. These endemic species, found nowhere else in the world, have a variety of origins. Some evolved during the Holocene as modern coastal plain communities assembled and disturbance regimes characterized by frequent fires and hurricanes became prevalent. Others represent more ancient lineages and must have survived on the coastal plains throughout the many climate oscillations of the Quaternary. Included in this group are species local to Blufflands-style habitats along the Apalachicola River, boggy wetlands in the Carolinas and Georgia, and sparsely vegetated sandhills of the northern panhandle of Florida. As the climate continues to warm, and as fire frequency and hurricane occurrence increase, it is possible that these species will survive and may even expand their ranges northward unless threatened by habitat loss due to encroachment of human retirement communities or unless out-competed by introduced exotic species.

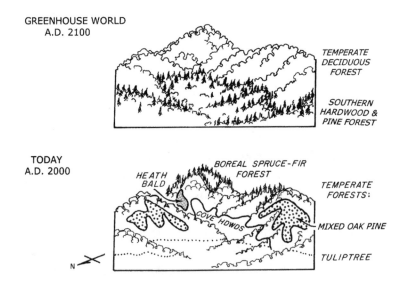

Figure 34. Future vegetation patterns on Mount LeConte, Great Smoky Mountains National Park, projected for the year AD 2100 based on data shown in Figure 33.

To me, studying the past is important for gaining an appreciation of the dynamics of ecological pattern and process that have led to development of the landscapes of the present, and for the implications such studies have for conserving biological diversity into the future. The Quaternary record of long-term changes in climate and vegetation in the eastern deciduous forest region has shown me how important it is to protect remaining populations of rare species. But a conservation goal that is even more important is to establish routes for their escape — corridors for migration that will ensure survival of temperate forest species through range adjustments that will have to be made if they are to survive global warming.

Epilogue:

An Ecofuturist's View of the
Eastern Deciduous Forest

My professional journey, beginning as plant geographer, becoming historian of trees, then reconstructing past landscapes as Quaternary paleoecologist, has come full circle. I now consider myself to be ecofuturist as well. By gaining an appreciation of the past, I realize how important it is also to be forward thinking.

Early in the 20th century, ecologists had largely a static view of plant communities. Because life zones seemed fixed in latitude and altitude on the modern landscape, to plant geographers they represented a nature that could be understood through description and mapping. Old-growth forests of eastern North America were seen as primeval, virgin wilderness in which only the balance of nature was to be found, without interference by human alteration. This romanticized notion of "natural" vegetation persists in the form of federal policy regarding restoration of damaged ecosystems. We still seek ways to cause ecosystems to revert to an idealized pre-existing condition that somehow was better than any that can be found on today's fragmented landscape.

As a historian of trees, I learned to think in "tree time." Along with other plant ecologists of the late 20th century and early 21st century, I view vegetation as a dynamic entity. Genes flow from one generation of trees to the next as pollen and seeds are dispersed, and populations consisting of many individuals grow and decline in response to changes in environment. The changes are keyed to differences in life history characteristics of the species and their tolerances and competitive abilities. Communities change with time because of the dynamic interactions of species across complex landscape mosaics.

The paleoecological perspective adds another layer of complexity to my perception of the dynamic nature of vegetation. Studies of specific landscape histories make it clear that changes in vegetation occur on all scales in space and time. Dramatic changes in vegetation have occurred in response to Quaternary climate change. Humans also have played a fundamental role in shaping the distribution and composition of biological communities throughout the Holocene. The so-called PreColumbian time line of five hundred years ago (before Christopher Columbus arrived in the New World) used as a base line for change reflects only the historic period of most recent colonization and human disturbance of the eastern deciduous forest region. It is a convenient but arbitrary and Euro-centric basis upon which to judge the trajectory of change in vegetation.

Change in climate and biological communities is a reality brought home by the many studies of long-term vegetation history amassed by Quaternary paleoecologists. The prospect of near-future global climate warming is becoming an accepted outcome of human activities as a result of the continued use of fossil fuels. In the 21st century and beyond, not only is the rate of climate change accelerating beyond that documented in the Quaternary record, but also the magnitude of climate warming will probably be greater than that of any previous interglacial interval. In addition, the future Holocene "super-interglacial" may last for the next fifty thousand years, a very long time frame for either humans or trees.

The implications of the past for the future of the eastern deciduous forest and other imperiled sectors of the biota are clear. It is a false goal to think that we can and should transform our landscapes into ones resembling a PreColumbian primeval wilderness. Five hundred years ago, Northern Hemisphere climate was in the midst of a relatively cold interval called the "Little Ice Age" that is not likely to be repeated in the near future. The vegetation that European pioneer settlers encountered then was not wilderness untouched by human activities. It was a dynamic mosaic of species still adjusting to late Holocene climate change and recovering from the influence of prehistoric human disturbance, including use of fire.

As an ecofuturist, I am optimistic about the fate of the eastern deciduous forest. I firmly believe that it is essential to understand as much as we can about the history of such a natural system in order to preserve it for the future. But instead of living in the past, hoping to restore ecosystems to some state that existed under conditions that cannot now be duplicated, it is important to look forward and to use our understanding of how the fundamental processes underlying changes in biological communities will affect their continued persistence. Using our knowledge of the physiological tolerances and life-history characteristics of the species, along with our understanding of their responses to changes in climate and other aspects of environment, we can project their potential vulnerability to global warming and its associated environmental changes in time to make a difference in the future survival of imperiled species. Through the means of scientific investigation that we have at hand, *it is within our ability* to understand and to act in responsible ways in order to ensure the preservation of a great many species that are now potentially vulnerable to extinction. We may elect to find indirect ways to preserve biological diversity by setting aside lands to act as corridors for migration and habitat expansion or relocation. On the other hand, we may choose to manipulate community composition directly through removal of unwanted exotic species and replanting or seeding in desirable native species into potentially habitable areas. Which alternative is more appropriate is the kind of ethical question conservationists need to ask. The future of our imperiled eastern deciduous forest depends largely upon how such choices are made. These decisions require thinking ahead and allowing for change rather than trying to recreate the past.

References

Berry, E. W. 1916. The Lower Eocene floras of southeastern North America. *United States Geological Survey Professional Paper* 91: 1-481.

Berry, E. W. 1924. The Middle and Upper Eocene floras of southeastern North America. *United States Geological Survey Professional Paper* 92: 1-206.

Bourdo, E. A., Jr. 1956. A review of the General Land Survey and its use in quantitative studies of former forests. *Ecology* 37: 754-768.

Brain, J. 1977. On the Tunica trail. *Louisiana Archaeological Survey and Antiquities Commission Anthropological Study No. 1*. Baton Rouge, LA: Department of Culture, Recreation and Tourism.

Braun, E. L. 1950. *Deciduous forests of eastern North America*. New York, NY: Hafner.

Braun, E. L. 1951. Plant distribution in relation to the glacial boundary. *Ohio Journal of Science* 51:139-146.

Braun, E. L. 1955. The phytogeography of unglaciated eastern United States and its interpretation. *Botanical Review* 21: 297-375.

Brouillet, L., and R. D. Whetstone. 1993. Climate and physiography, pp. 15-46 *in* N. R. Morin (ed.), *Flora of North America north of Mexico, Volume I, Introduction*. New York, NY: Oxford University Press.

Brown, C. A. 1938. The flora of Pleistocene deposits in the western Florida Parishes, West Feliciana Parish, and East Baton Rouge Parish, Louisiana. *Louisiana Department of Conservation Geological Bulletin* 12: 59-96.

Brunner, C. A. 1982. Paleoceanography of surface waters in the Gulf of Mexico during the late Quaternary. *Quaternary Research* 17: 105-119.

Bryant, V. M. Jr., and R. G. Holloway. 1985. A late-Quaternary paleoenvironmental record of Texas: An overview of the pollen evidence, pp. 39-70 *in* V. M. Bryant, Jr., and R. Holloway (eds.), *Pollen records of late-Quaternary North American sediments*. Dallas, TX: American Association of Stratigraphic Palynologists.

Buell, M. F. 1945. Late Pleistocene forests of southeastern North Carolina. *Torreya* 45: 117-118.

Bull, J., and J. Farrand, Jr. 1977. *The Audubon Society field guide to North American birds.* New York, NY: Alfred A. Knopf.

Cain, S. A. 1943. The Tertiary nature of the cove hardwood forests of the Great Smoky Mountains. *Bulletin of the Torrey Botanical Club* 70: 213-235.

Cain, S. A.1944. *Foundations of plant geography.* New York, NY: Harper & Brothers.

Chaney, R. W. 1947. Tertiary centers and migration routes. *Ecological Monographs* 17: 140-148.

Chapman, J. 1994. *Tellico archaeology: 12,000 years of Native American history (revised edition).* Knoxville, TN: University of Tennessee Press.

Clark, J. S., C. Fastie, G. Hurtt, S. T. Jackson, C. Johnson, G. A. King, M. Lewis, J. Lynch, S. Pacala, C. Prentice, E. W. Schupp, T. Webb III, and P. Wyckoff. 1998. Reid's paradox of rapid plant migration. *BioScience* 48: 13-24.

Cocks, R. S. 1914. Notes on the flora of Louisiana. *The Plant World* 17: 186-191.

Corning, H. (ed.) 1929. *Journal of John James Audubon made during his trip to New Orleans in 1820-1821.* Boston, MA: Club of Odd Volumes.

Cridlebaugh, P. A. 1984. *American Indian and Euro-American impact upon Holocene vegetation in the Lower Little Tennessee River Valley, East Tennessee.* Dissertation, Department of Anthropology, University of Tennessee, Knoxville.

Cronon, W. 1983. *Changes in the land: Indians, colonists, and the ecology of New England.* New York, NY: Hill and Wang.

Davidson, J. L. 1983. *Paleoecological analysis of Holocene vegetation, Lake in the Woods, Cades Cove, Great Smoky Mountains National Park.* Thesis, Graduate Progam in Ecology, University of Tennessee, Knoxville.

Davis, J. H., Jr. 1946. The peat deposits of Florida, their occurrence, development, and uses. *Florida Department of Conservation, Geological Survey, Geological Bulletin* 30: 1-247.

Davis, M. B. 1976. Pleistocene biogeography of temperate deciduous forests. *Geoscience and Man* 113: 13-26.

Davis, M. B. 1983. Quaternary history of deciduous forests of eastern North America and Europe. *Annals of the Missouri Botanical Garden* 70: 550-563.

Davis, M. B. 1986. Climatic instability, time lags, and community disequilibrium, pp. 269-284 *in* J. Diamond and T. J. Case (eds.), *Community ecology.* New York, NY: Harper and Row.

Davis, M. B., and C. Zabinski. 1992. Changes in geographical range resulting from greenhouse warming: Effects on biodiversity in forests, pp. 297-308 *in* R. L. Peters and T. E. Lovejoy (eds.), *Global Warming and Biological Diversity*. New Haven, CT: Yale University Press.

Deevey, E. S., Jr. 1949. Biogeography of the Pleistocene, Part I: Europe and North America. *Bulletin of the Geological Society of America* 60: 1315-1416.

Delcourt, H. R. 1974. *Late Quaternary history of the mixed mesophytic forest in Mississippi and Louisiana.* Thesis, Department of Botany, Louisiana State University, Baton Rouge.

Delcourt, H. R. 1975. Reconstructing the forest primeval, West Feliciana Parish, Louisiana. *Louisiana State University Museum of Geosciences, Mélanges Series* 10: 1-13.

Delcourt, H. R. 1978. *Late Quaternary vegetation history of the eastern Highland Rim and adjacent Cumberland Plateau of Tennessee.* Dissertation, Department of Ecology and Behavioral Biology, University of Minnesota, Minneapolis.

Delcourt, H. R. 1979. Late Quaternary vegetation history of the eastern Highland Rim and adjacent Cumberland Plateau of Tennessee. *Ecological Monographs* 49: 255-280.

Delcourt, H. R. 1981. The virtue of forests, virgin and otherwise. *Natural History* 90(6): 32-39.

Delcourt, H. R. 1985. Holocene vegetational changes in the southern Appalachian Mountains, U.S.A. *Ecologia Mediterranea* 11: 9-16.

Delcourt, H. R. 1987. The impact of prehistoric agriculture and land occupation on natural vegetation. *Trends in Ecology and Evolution* 2: 39-44.

Delcourt, H. R., and P. A. Delcourt. 1974. Primeval magnolia-holly-beech climax in Louisiana. *Ecology* 55: 638-644.

Delcourt, H. R., and P. A. Delcourt. 1975. The Blufflands: Pleistocene pathway into the Tunica Hills. *American Midland Naturalist* 94: 385-400.

Delcourt, H. R., and P. A. Delcourt. 1977. Presettlement magnolia-beech climax of the Gulf Coastal Plain: Quantitative evidence from the Apalachicola River bluffs, north-central Florida. *Ecology* 58: 1085-1093.

Delcourt, H. R., and P. A. Delcourt. 1984. Ice age haven for hardwoods. *Natural History* 93(9): 22-28.

Delcourt, H. R., and P. A. Delcourt. 1985. Quaternary palynology and vegetational history of the southeastern United States, pp. 1-37 *in* V. M. Bryant, Jr., and R. Holloway (eds.), *Pollen records of late-Quaternary North American sediments*. Dallas, TX: American Association of Stratigraphic Palynologists.

Delcourt, H. R., and P. A. Delcourt. 1986. Late-Quaternary vegetational history of the central Atlantic states, pp. 23-35 *in* J. N. McDonald and S. O. Bird (eds.), *The Quaternary of Virginia*. Virginia Commonwealth Division of Mineral Resources.

Delcourt, H. R., and P. A. Delcourt. 1988. Quaternary landscape ecology: Relevant scales in space and time. *Landscape Ecology* 2: 23-44.

Delcourt, H. R., and P. A. Delcourt. 1991. *Quaternary ecology, a paleoecological perspective*. New York NY: Chapman and Hall.

Delcourt, H. R., and P. A. Delcourt. 1991. Late-Quaternary vegetation history of the Interior Highland region of Missouri, Arkansas, and Oklahoma, pp. 15-30 *in* L. Hedrick (ed.), *Restoring old-growth forests in the interior highlands of Arkansas and Oklahoma*. Morrilton, AR: United States Department of Agriculture Forest Service and Winrock International.

Delcourt, H. R., and P. A. Delcourt. 1994. Postglacial rise and decline of *Ostrya virginiana* (Mill.) K. Koch and *Carpinus caroliniana* Walt. in eastern North America: Predictable responses of forest species to cyclic changes in seasonality of climates. *Journal of Biogeography* 21: 137-150.

Delcourt, H. R., and P. A. Delcourt. 2000. Eastern deciduous forests, Chapter 10, pp. 357-396 *in* M. G. Barbour and W. D. Billings (eds.), *North American terrestrial vegetation, 2nd edition*. Cambridge, UK: Cambridge University Press.

Delcourt, H. R., P. A. Delcourt, and P. D. Royall. 1997. Late-Quaternary vegetational history of the Western Lowlands, pp. 103-122 *in* D. F. Morse (ed.), *Sloan, a Paleoindian/Dalton period cemetery in Arkansas*. Washington, D. C.: Smithsonian Institution Press.

Delcourt, P. A. 1974. *Quaternary geology and paleoecology of West and East Feliciana Parishes, Louisiana, and Wilkinson County, Mississippi*. Thesis, Department of Geosciences, Louisiana State University, Baton Rouge.

Delcourt, P. A. 1978. *Quaternary history of the Gulf Coastal Plain*. Dissertation, Department of Geology, University of Minnesota, Minneapolis.

Delcourt, P. A. 1980. Goshen Springs: Late Quaternary vegetation record for southern Alabama. *Ecology* 61: 371-386.

Delcourt, P. A., and H. R. Delcourt. 1977. The Tunica Hills, Louisiana-Mississippi: Late glacial locality for spruce and deciduous forest species. *Quaternary Research* 7: 218-237.

Delcourt, P. A., and H. R. Delcourt. 1978. Discussion of "The Tunica Hills, Louisiana-Mississippi: Late glacial locality for spruce and deciduous forest species." *Quaternary Research* 9: 253-259.

Delcourt, P. A., and H. R. Delcourt. 1979. Late Pleistocene and Holocene distributional history of the deciduous forest in the southeastern United States. *Veröffentilichungen des Geobotanischen Institutes der ETH, Stiftung Rübel (Zurich)* 68: 79-107.

Delcourt, P. A., and H. R. Delcourt. 1981. Vegetation maps for eastern North America: 40,000 yr BP to the present, pp. 123-165 *in* R. C. Romans (ed.), *Geobotany II*. New York, NY: Plenum Press.

Delcourt, P. A., and H. R. Delcourt. 1982. Invited Research Feature: Quaternary paleoecology of the southeastern United States. *Bulletin of the Association of Southeastern Biologists* 29: 135-138.

Delcourt, P. A., and H. R. Delcourt. 1983. Late-Quaternary vegetation dynamics and community stability reconsidered. *Quaternary Research* 19: 265-271.

Delcourt, P. A., and H. R. Delcourt. 1984. Late-Quaternary paleoclimates and biotic responses in eastern North America and the western North Atlantic Ocean. *Palaeogeography, Palaeoclimatology, Palaeoecology* 48: 263-284.

Delcourt, P. A., and H. R. Delcourt. 1985. Dynamic landscapes of East Tennessee: An integration of paleoecology, geomorphology, and archaeology. *University of Tennessee, Knoxville, Department of Geological Sciences, Studies in Geology* 9: 191-220.

Delcourt, P. A., and H. R. Delcourt. 1987. *Long-term forest dynamics of the Temperate Zone, a case study of late-Quaternary forests in eastern North America, Ecological Studies 63*. New York, NY: Springer-Verlag.

Delcourt, P. A., and H. R. Delcourt. 1987. Late-Quaternary dynamics of temperate forests: Applications of paleoecology to issues of global environmental change. *Quaternary Science Reviews* 6: 129-146.

Delcourt, P. A., and H. R. Delcourt. 1992. Ecotone dynamics in space and time, pp. 19-54 *in* A. J. Hansen and F. di Castri (eds.), *Landscape boundaries, consequences for biotic diversity and ecological flows*. New York, NY: Springer-Verlag.

Delcourt, P. A., and H. R. Delcourt. 1993. Paleoclimates, paleovegetation, and paleofloras of North America during the late Quaternary, pp. 71-94 *in* N. R. Morin and L. Brouillet (eds.), *Flora of North America north of Mexico, Volume I*. Oxford, UK: Oxford University Press,

Delcourt, P. A., and H. R. Delcourt. 1996. Quaternary vegetation history of the Lower Mississippi Valley. *Engineering Geology* 45: 219-242.

Delcourt, P. A., and H. R. Delcourt. 1998. Paleoecological insights on conservation of biodiversity: A focus on species, ecosystems, and landscapes. *Ecological Applications* 8: 921-934.

Delcourt, P. A., and H. R. Delcourt. 2002. *Prehistoric Native Americans and ecological change.* Cambridge, UK: Cambridge University Press.

Delcourt, P. A., O. K. Davis, and R. C. Bright. 1979. Bibliography of taxonomic literature for the identification of fruits, seeds, and vegetative plant fragments.*Oak Ridge National Laboratory, Oak Ridge, Tennessee, ORNL/TM*-6818: 1-84.

Delcourt, P. A., H. R. Delcourt, and J. L. Davidson. 1983. Mapping and calibration of modern pollen-vegetation relationships in the southeastern United States. *Review of Palaeobotany and Palynology* 39: 1-45.

Delcourt, P. A., H. R. Delcourt, and R. T. Saucier. 1999. Late-Quaternary vegetational dynamics in the Central Mississippi Valley, pp. 15-30 *in* R. C. Mainfort, Jr., and M. D. Jeter (eds.), *Arkansas Archaeology.* Fayetteville, AR: University of Arkansas Press.

Delcourt, P. A., H. R. Delcourt, and T. Webb, III. 1984. Atlas of mapped distributions of dominance and modern pollen percentages for important tree taxa of eastern North America. *American Association of Stratigraphic Palynologists, Contributions Series* 14: 1-131.

Delcourt, P. A., H. R. Delcourt, R. C. Brister, and L. E. Lackey. 1980. Quaternary vegetation history of the Mississippi Embayment. *Quaternary Research* 13: 111-132.

Delcourt, P. A., H. R. Delcourt, C. R. Ison, W. E. Sharp, and K. J. Gremillion. 1998. Prehistoric human use of fire, the eastern agricultural complex, and Appalachian oak-chestnut forests: Paleoecology of Cliff Palace Pond, Kentucky. *American Antiquity* 63: 263-278.

Delcourt, P. A., H. R. Delcourt, D. F. Morse, and P. A. Morse. 1993. History, evolution, and organization of vegetation and human culture, pp. 47-79 *in* W. H. Martin, S. G. Boyce, and A. C. Echternacht (eds.), *Biodiversity of the southeastern United States: Lowland terrestrial communities.* New York, NY: John Wiley and Sons.

Dixon, E. J. 1999. *Bones, boats, & bison, archeology and the first colonization of western North America.* Albuquerque, NM: University of New Mexico Press.

Faegri, K., and J. Iversen. 1975. *Textbook of pollen analysis, 3rd edition.* New York, NY: Hafner.

Fernald, M. L. 1950. *Gray's manual of botany, 8th ed.* New York, NY: American Book Company.

Fisk, H. N. 1938. Pleistocene exposures in western Florida Parishes, Louisiana. *Louisiana Department of Conservation Geological Bulletin* 12: 3-25.

Fowells, H. A. 1965. Silvics of forest trees of the United States. *United States Forest Service Agricultural Handbook* 271: 1-762.

Frey, D. G. 1951. Pollen succession in the sediments of Singletary Lake, North Carolina. *Ecology* 32: 518-533.

Gardner, J. S., and C. Ettingshausen. 1879. *A monograph of the British Eocene flora.* London, UK: London Palaeontographical Society.

Gleason, H. A. 1926. The individualistic concept of the plant association. *Bulletin of the Torrey Botanical Club* 53:7-26.

Graham, A. (ed.) 1972. Floristics and paleofloristics of Asia and eastern North America. Amsterdam, Netherlands: Elsevier.

Grüger, E. 1972. Late Quaternary vegetation development in south-central Illinois. *Quaternary Research* 2: 217-231.

Harcombe, P. A., and P. L. Marks. 1978. Tree diameter distributions and replacement processes in southeast Texas forests. *Forest Science* 24: 153-166.

Henry, J. 1858. Meteorology and its connection with agriculture, pp. 429-493 *in* U.S. Congress, House *Report of the Commissioner of Patents for 1858, 35th Congress, 1st session, 1858, H. Ex. Doc.* 32 (serial no. 954).

Hilgard, E. W. 1860. *Report on the geology and agriculture of the state of Mississippi.* Jackson, MS: E. Barksdale, State Printer.

Hills, L. V., and R. T. Ogilvie. 1970. *Picea banksii* n. sp. Beaufort Formation (Tertiary), northwestern Banks Island, Arctic Canada. *Canadian Journal of Botany* 48: 457-464.

Hudson, C. 1997. *Knights of Spain, Warriors of the Sun: Hernando de Soto and the South's ancient chiefdoms.* Athens, GA: University of Georgia Press.

Humboldt, A. von. 1817. *De distributione geographica plantarum, secundum cocli temperiem et altitudinem montium prolegomena.* Paris, France: Chez Levrault, Schoell et Compagne.

Imbrie, J., and K. P. Imbrie. 1979. *Ice ages: Solving the mystery.* Hillside, NJ: Enslow Publishers.

Iverson, L. R., and A. M. Prasad. 1998. Predicting abundance of 80 tree species following climate change in the eastern United States. *Ecological Monographs* 68: 465-485.

Iverson, L. R., A. M. Prasad, B. J. Hale, and E. K. Sutherland. 1999. An atlas of current and potential future distributions of common trees of the eastern United States. *Northeastern Research Station, USDA Forest Service, Delaware, Ohio, General Technical Report* NE-265.

Jackson, S. T., and C. R. Givens. 1994. Late Wisconsinan vegetation and environment of the Tunica Hills region, Louisiana/Mississippi. *Quaternary Research* 41: 316-325.

Jackson, S. T., and C. Weng. 1999. Late Quaternary extinction of a tree species in eastern North America. *Proceedings of the National Academy of Sciences* 96: 13847-13852.

Kaczorowski, R. T. 1977. The Carolina Bays: A comparison with modern oriented lakes. *Coastal Research Division, Department of Geology, University of South Carolina, Columbia, Technical Report* 13-CRD: 1-124.

Kapp, R. O. 1968. *How to know pollen and spores.* Dubuque, IA: Wm. C. Brown Company.

Kim, K.-Y., and T. J. Crowley. 1994. Modeling the climate effect of unrestricted Greenhouse emissions over the next 10,000 years. *Geophysical Research Letters* 21: 681-684.

King, J. E. 1973. Late Pleistocene palynology and biogeography of the western Missouri Ozarks. *Ecological Monographs* 43: 539-565.

Krech, S. III. 1999. *The ecological Indian, myth and history.* New York, NY: W. W. Norton & Company.

Krinitzsky, E. L., and W. J. Turnbull. 1967. Loess deposits of Mississippi. *Geological Society of America Special Paper* 94: 1-64.

Kurtén, B., and E. Anderson. 1980. *Pleistocene mammals of North America.* New York, NY: Columbia University Press.

Larabee, P. A. 1986. *Late-Quaternary vegetational and geomorphic history of the Allegheny Plateau at Big Run Bog, Tucker County, West Virginia.* Thesis, Department of Geological Sciences, University of Tennessee, Knoxville.

Loutre, M. F., and A. Berger. 2000. Future climatic changes: Are we entering an exceptionally long interglacial? *Climatic Change* 46: 61-90.

Lyell, C. 1847. On the delta and alluvial deposits of the Mississippi River, and other points in the geology of North America, observed in the years 1845, 1846. *American Journal of Science* 3: 34-39, 267-269.

Martin, P. S. 1958. Taiga-tundra and the full-glacial period in Chester County, Pennsylvania. *American Journal of Science* 256: 470-502.

Martin, P. S. 1958. Pleistocene ecology and biogeography of North America, pp. 375-420 *in* C. S. Hubbs (ed.), *Zoogeography.* American Association for the Advancement of Science Publication 51.

Martin, P. S., and B. E. Harrell. 1957. The Pleistocene history of temperate biotas in Mexico and eastern United States. *Ecology* 38: 468-480.

Martin, P. S., and R. G. Klein. 1984. *Quaternary extinctions, a prehistoric revolution.* Tucson, AZ: University of Arizona Press.

Martin, P. S., and P. J. Mehringer, Jr. 1965. Pleistocene pollen analysis and biogeography of the Southwest, pp. 433-451 *in* H. E. Wright, Jr., and D. G. Frey (eds.), *The Quaternary of the United States.* Princeton, NJ: Princeton University Press.

Martin, P. S., and H. E. Wright, Jr. (eds.) 1967. *Pleistocene extinctions: The search for a cause.* New Haven, CT: Yale University Press.

Maxwell, J. A., and M. B. Davis. 1972. Pollen evidence of Pleistocene and Holocene vegetation on the Allegheny Plateau, Maryland. *Quaternary Research* 2: 506-530.

McAndrews, J. H., A. A. Berti, and G. Norris. 1973. *Key to the Quaternary pollen and spores of the Great Lakes region*. Toronto, ONT: Royal Ontario Museum.

McDonald, J. N., and C. S. Bartlett, Jr. 1983. An associated musk ox skeleton from Saltville, Virginia. *Journal of Vertebrate Paleontology* 2: 453-470.

Milankovitch, M. M. 1941. Canon of insolation and the ice-age problem. *Royal Serbian Academy Special Publication* 133.

Morse, D. F., and Morse, P. A. 1983. *Archaeology of the Central Mississippi Valley*. New York, NY: Academic Press.

Murrell, Z. E. 2001. Preface, southeastern endemics: Speciation and biogeography. *Castanea* 66: 1-2.

Newton, M. B., Jr. 1972. Atlas of Louisiana. *Louisiana State University, School of Geosciences Miscellaneous Publication* 72-1.

Olsson, I. U. 1986. Radiometric dating, pp. 273-312 *in* B. E. Berglund (ed.), *Handbook of Holocene palaeoecology and palaeohydrology*. Chichester, UK: John Wiley and Sons.

Pielou, E. C. 1991. *After the Ice Age: The return of life to glaciated North America*. Chicago, IL: University of Chicago Press.

Potzger, J. E., and B. C. Tharp. 1954. Pollen study of two bogs in Texas. *Ecology* 35: 462-466.

Price, J. E. 1978. The settlement pattern of the Powers Phase, pp. 201-231 *in* B. D. Smith (ed.), *Mississippian settlement patterns*. New York, NY: Academic Press.

Quarterman, E. 1981. A fresh look at climax forests of the Coastal Plain. *Bulletin of the Association of Southeastern Biologists* 28: 143-148.

Quarterman, E., and C. Keever. 1962. Southern mixed hardwood forest: Climax in the southeast coastal plain, U.S.A. *Ecological Monographs* 32: 167-185.

Redman, C. L. 1999. *Human impact on ancient environments*. Tucson, AZ: University of Arizona Press.

Rowland, D. 1930. *Life, letters, and papers of William Dunbar of Elgin, Morrayshire, Scotland, and Natchez (1749-1810)*. Jackson, MS: Press of the Mississippi Historical Society.

Royall, P. D. 1988. *Late-Quaternary paleoecology and paleoenvironments of the Western Lowlands, southeast Missouri*. Thesis, Department of Geological Sciences, University of Tennessee, Knoxville.

Royall, P. D., P. A. Delcourt, and H. R. Delcourt. 1991. Late Quaternary paleoecology and paleoenvironments of the Central Mississippi Alluvial Valley. *Bulletin of the Geological Society of America* 103: 157-170.

Ruddiman, W. F., and H. E. Wright, Jr. (eds.) 1987. *North America and adjacent oceans during the last deglaciation, Vol. K-3, The Geology of North America.* Boulder, CO: Geological Society of America.

Russell, E. W. B. 1997. *People and the land through time: Linking ecology and history.* New Haven, CT: Yale University Press.

Sears, P. B., and K. H. Clisby. 1955. Palynology of southern North America. IV: Pleistocene climate of Mexico. *Bulletin of the Geological Society of America* 66: 21-530.

Shafer, D. S. 1984. *Late-Quaternary paleoecologic, geomorphic, and paleoclimatic history of Flat Laurel Gap, Blue Ridge Mountains, North Carolina.* Thesis, Department of Geological Sciences, University of Tennessee, Knoxville.

Shafer, D. S. 1985. Flat Laurel Gap Bog, Pisgah Ridge, North Carolina: Late Holocene development of a high-elevation heath bald. *Castanea* 51: 1-10.

Shafer, D. S. 1988. Late Quaternary landscape evolution at Flat Laurel Gap, Blue Ridge Mountains, North Carolina. *Quaternary Research* 30: 7-11.

Shantz, H. L., and R. Zon. 1924. The natural vegetation of the United States, pp. 1-29 *in* O. E. Baker, compiler. *Atlas of American agriculture, 1936.* Washington, D. C.: U. S. Department of Agriculture.

Sharp, A. J. 1951. The relation of the Eocene Wilcox Flora to some modern floras. *Evolution* 5: 1-5.

Smith, B. D. 1992. *Rivers of change: Essays on early agriculture in eastern North America.* Washington, D.C.: Smithsonian Institution Press.

Smith, E. N., Jr. 1984. *Late-Quaternary vegetational history at Cupola Pond, Ozark National Scenic Riverways, southeastern Missouri.* Thesis, Department of Geological Sciences, University of Tennessee, Knoxville.

Snowden, J. O., Jr., and R. R. Priddy. 1968. Loess investigations of Mississippi: Geology of Mississippi loess. *Mississippi Geological, Economic and Topographical Survey Bulletin* 111: 5-203.

Steyermark, J. A. 1959. Vegetational history of the Ozark forest. *The University of Missouri Studies* 31: 1-138.

Thompson, R. S., S. W. Hostetler, P. J. Bartlein, and K. H. Anderson. 1998. A strategy for assessing potential future changes in climate, hydrology, and vegetation in the western United States. *United States Geological Survey Circular* 1153.

Tolliver, R. 1998. *Diatom succession across the Pleistocene/Holocene interval: An example from the southeastern United States.* Dissertation, Department of Geological Sciences, University of Tennessee, Knoxville.

Van Doren, M. (ed.) 1928 (reprinted in 1955). *Travels of William Bartram.* New York, NY: Dover Publications.

Van Kat, J. L. 1979. *The natural vegetation of North America.* New York, NY: John Wiley & Sons.

VEMAP Members. 1995. Vegetation/Modeling and Analysis Project (VEMAP): Comparing biogeography and biogeochemistry models in a continental-scale study of terrestrial ecosystem responses to climate change and CO_2 doubling. *Global Biogeochemical Cycles* 9: 407-437.

Von Post, L. (1916, translated into English by M. B. Davis and K. Faegri, 1967). Forest tree pollen in south Swedish peat bog deposits. *Pollen et Spores* 9: 378-401.

Walter, H. 1979. *Vegetation of the Earth and ecological systems of the geobiosphere, 2nd ed.* New York, NY: Springer-Verlag.

Ware, S., C. Frost, and P. D. Doerr. 1993. Southern mixed hardwood forest: The former longleaf pine forest, pp. 447-493 *in* W. H. Martin, S. G. Boyce, and A. C. Echternacht (eds.), *Biodiversity of the southeastern United States: Lowland terrestrial communities.* New York, NY: John Wiley and Sons.

Watts, W. A. 1970. The full-glacial vegetation of northwestern Georgia. *Ecology* 51: 17-33.

Watts, W. A. 1975. A late Quaternary record of vegetation at Lake Annie, south-central Florida. *Geology* 3: 344-346.

Watts, W. A. 1979. Late Quaternary vegetation of central Appalachia and the New Jersey Coastal Plain. *Ecological Monographs* 49: 427-469.

Watts, W. A. 1980. Late-Quaternary vegetation history at White Pond on the inner Coastal Plain of South Carolina. *Quaternary Research* 13: 187-199.

Watts, W. A. 1983. Vegetational history of the eastern United States 25,000 to 10,000 years ago, pp. 294-310 *in* S. C. Porter (ed.), *Late-Quaternary environments of the United States, Volume I, the late Pleistocene.* Minneapolis, MN: University of Minnesota Press.

Watts, W. A., and J. P. Bradbury. 1982. Paleoecological studies at Lake Patzcuaro on the west-central Mexican Plateau and at Chalco in the Basin of Mexico. *Quaternary Research* 17: 56-70.

Watts, W. A., B. C. S. Hansen, and E. C. Grimm. 1992. Camel Lake: A 40,000-YR record of vegetational and forest history from Northwest Florida. *Ecology* 73: 1056-1066.

Watts, W. A., and M. Stuiver. 1980. Late Wisconsin climate of northern Florida and the origin of species-rich deciduous forest. *Science* 210: 325-327.

Webb, S. L. 1986. Potential role of passenger pigeons and other vertebrates in the rapid Holocene migrations of nut trees. *Quaternary Research* 26: 367-375.

Whitehead, D. R. 1964. Fossil pine pollen and full-glacial vegetation in southeastern North Carolina. *Ecology* 45: 767-777.

Whitehead, D. R. 1973. Late-Wisconsin vegetational changes in unglaciated eastern North America. *Quaternary Research* 3: 621-631.

Whitehead, D. R. 1981. Late-Pleistocene vegetational changes in northeastern North Carolina. *Ecological Monographs* 51:451-471.

Whitney, G. G. 1994. *From coastal wilderness to fruited plain: A history of environmental change in temperate North America from 1500 to the present.* Cambridge, UK: Cambridge University Press.

Whittaker, R. H. 1975. *Communities and ecosystems, 2nd ed.* New York, NY: MacMillan.

Wilkins, G. R. 1985. *Late-Quaternary vegetational history at Jackson Pond, Larue County, Kentucky.* Thesis, Department of Geological Sciences, University of Tennessee, Knoxville.

Wilkins, G. R., P. A. Delcourt, H. R. Delcourt, F. W. Harrison, and M. R. Turner. 1991. Paleoecology of central Kentucky since the last glacial maximum. *Quaternary Research* 36: 224-239.

Williams, M. 1989. *Americans and their forests, a historical geography.* Cambridge, UK: Cambridge University Press.

Wright, H. E., Jr. 1967. A square-rod piston sampler for lake sediments. *Journal of Sedimentary Petrology* 37: 975-976.

Wright, H. E., Jr. 1968. History of the Prairie Peninsula, pp. 78-88 *in* R. E. Bergstrom (ed.), *The Quaternary of Illinois, Special Report* 14. Urbana, IL: University of Illinois College of Agriculture.

Wright, H. E., Jr. 1976. Pleistocene ecology — some current problems. *Geoscience and Man* 13: 1-12.

Wright, H. E., Jr. 1976. The dynamic nature of Holocene vegetation, a problem in paleoclimatology, biogeography, and stratigraphic nomenclature. *Quaternary Research* 6: 581-596.

Wright, H. E., Jr. 1977. Quaternary vegetation history — some comparisons between Europe and America. *Annual Review of Earth and Planetary Sciences* 5: 123-158.

Wright, H. E., Jr. 1984. Sensitivity and response time of natural systems to climatic change in the late Quaternary. *Quaternary Science Reviews* 3: 91-131.

Subject Index

Index of Common and Scientific Names

232